A YEAR WITH THE SEALS

A YEAR WITH THE SEALS

*Unlocking the Secrets
of the Sea's Most Charismatic
and Controversial Creatures*

ALIX MORRIS

ITHAKA

First published in the UK in 2025 by Ithaka Press
An imprint of Bonnier Books UK
5th Floor, HYLO, 105 Bunhill Row,
London, EC1Y 8LZ

Copyright © Alix Morris, 2025

All rights reserved.
No part of this publication may be reproduced, stored or transmitted in any form or by any means, electronic, mechanical, photocopying or otherwise, without the prior written permission of the publisher.

The right of Alix Morris to be identified as Author of this work has been asserted by her in accordance with the Copyright, Designs and Patents Act, 1988.

A CIP catalogue record for this book is available from the British Library.

Hardback ISBN: 978-1-78512-762-5

Also available as an ebook and an audiobook

1 3 5 7 9 10 8 6 4 2

Printed and bound in Great Britain by Clays Ltd, Elcograf S.p.A.

Every reasonable effort has been made to trace copyright holders of material reproduced in this book, but if any have been inadvertently overlooked the publishers would be glad to hear from them.

The authorised representative in the EEA is Bonnier Books
UK (Ireland) Limited.
Registered office address: Floor 3, Block 3, Miesian Plaza,
Dublin 2, D02 Y754, Ireland
compliance@bonnierbooks.ie
www.bonnierbooks.co.uk

*For my parents, who taught me
to be curious, and to care.*

CONTENTS

Foreword by Sy Montgomery   ix

Preface   xv

PART ONE: AUTUMN   1

Map of Northwest Atlantic   2

1: A Seal with Something to Say   3

2: The Traveling Seal   19

3: The Seal Bounty Conspiracy   34

4: The Mimic   47

PART TWO: WINTER   65

5: Why Did the Seal Cross the Road?   67

6: The Graveyard of the Atlantic   84

7: The Corkscrew Seal Mystery   98

8: Scamps or Scapegoats?   118

PART THREE: SPRING 133

Map of Puget Sound, Washington 134

9: Blubber Busters 135

10: The Fish Wars 151

11: The Chase 169

12: The Seal Snatcher 184

PART FOUR: SUMMER 195

13: Pinnipeds as Prey 197

14: Wild Cape Cod 215

15: War and Peacebuilding 228

16: The Shape-Shifter 243

Acknowledgments 255

Bibliography 259

FOREWORD
by Sy Montgomery

I SHOULD CONFESS TO YOU, dear reader, right from the start: I have always been pro-seal.

In reading this splendid, warm, and exhaustively researched book about these charismatic and controversial animals, I started out squarely on the side of the native animals.

After all, they were here first. The ancestors of today's modern seals and sea lions reached the west coast of North America about twenty-eight to thirty million years ago from the North Pacific. The first humans to arrive in North America were latecomers: they showed up a mere twenty-five thousand years ago. (And the immigrants who arrived on the *Mayflower*? They only landed in 1620.)

We humans, having evolved in Africa, are an invasive species. When we first showed up on this continent, who knows how many seals there were? But we do know there were plenty of fish for everyone back then. So who are we—having multiplied at the rate of shower mold to cover 90 percent of the earth with ourselves and our stuff—to complain, as some do today, that "too many" seals are "eating all the fish"? How can we blame the native wildlife when, on very rare occasions, a great white shark mistakes a human for its seal prey and injures or kills a person?

But I write this from my comfy home office in New Hampshire, where I make my living writing books—not fishing. Not surfing. Not earning my income from visitors to a formerly shark-free beach.

Another confession: though I've not eaten birds or mammals for forty years, I do sometimes enjoy wild-caught seafood. I seldom visit the beach, but I love it when I do, and I swim and scuba dive in the sea—where I would rather not be mistaken for a seal and eaten by a shark. And I deeply respect the life ways of Native Americans, and support their struggles in the face of centuries of persecution under the rule of the United States government—a government that I support with my taxes. That rule has included legal restrictions on hunting seals, both for food and to protect traditional fisheries Native people depend upon to survive.

It's impossible for me not to love seals, though. They are so doglike. In fact, they share ancestry with our dogs: the pinnipeds (seals, sea lions, and walruses) are all descendants of an ancient group called the Arctoid Carnivores—bears, dogs, and seals. Just look at their faces and you'll see the resemblance. And like dogs and bears, human-raised seals easily learn to enjoy our company (as you'll soon read, in the story of Andre the seal) and to emulate all sorts of human behaviors (you will learn about one seal, Hoover, who even learned to speak—with a Down East accent!).

Like the author of this book, my good friend Alix Morris, I have several times had the opportunity to be kissed by trained seals. (I liked it!) I admire their fluid agility in the water; I love watching them galumph on land; and I am constantly dazzled by their intelligence. In the 1980s, I visited a Seal Cognition Lab where researchers were successfully teaching seals and sea lions to use symbols to string together the equivalent of words into sentences. The animals were able to demonstrate to the scientists that they understood not just the meaning of the symbols, but that their position in the "sentence" is crucial. "Bring the ball to the stick" is different from "bring the stick to the ball." That is known in human language as syntax, and is recognized as a sign of sophisticated intelligence. They also understand musical rhythms, which demands a kind of pattern recognition known as beat perception and synchronization, another hallmark of advanced cognition.

Though seals may possess many of the mental abilities we admire in our own species, we are not always on the same wavelength. Years ago, also back in the eighties, I had the opportunity to work with a seal researcher who invited me to help tag gray seals she was studying in Maine. We were focused on young animals whose mothers had recently returned to the ocean, leaving the babies to finish growing their waterproof, adult coats before, themselves, heading out to sea. These babies, with their huge, liquid eyes and doglike faces, were so adorable that I worried that I might not be able to forcibly hold their head and shoulders down so the researcher could attach a plastic tag to a hind flipper, a quick and relatively painless process rather like piercing an ear. But doing this correctly, the researcher explained gravely, was essential so the seal would not bite us. And a seal bite is no joke. It can cause extensive bleeding. It can rip and tear flesh. It can break bones.

And then she told me about seal finger.

This is a serious and hideously painful infection that can afflict people who handle seals. The infected bite will swell up, hot and red—and there was no antibiotic that worked to cure it. Some patients begged docs to amputate the finger, because the pain was unbearable. (The bacterium responsible, a mycoplasma, was discovered only in 1991, and today it can be treated effectively.)

I promised my new friend I would hold the baby seal down. I started out attempting to do this gently—but then the horrified youngster, understandably desperate to escape, let out an unearthly bellow, twisted its incredibly strong, thick neck, and displayed to me some of its thirty-four, needle-sharp, backward-pointing teeth. I suddenly discovered I was quite capable of forcing the adorable baby's head down against the rock and holding it still, with strength I didn't know I had.

Yes, seals are adorable. Yes, seals are smart. But after her year with wild seals, Alix Morris tells us they are also strong, sometimes dangerous, wild animals, and that their agenda doesn't always fit with ours.

As I discovered when we worked on an Asiatic wild dog project together in Thailand, Alix Morris is an exceptionally skilled observer of wild animals. Her evocative, sometimes humorous, always compelling descriptions of seal behavior in this book are delightful and, alone, they are well worth the price of this book. But the pages you are about to read are about more than seals. Here you will read about seals *and people*. If you care about seals, you need to understand the source of the conflicts between them and our own kind—even if you, like me, are a seal supporter from the start.

Alix is a huge animal lover. But because she is an excellent journalist, she knows there is always more than one side to a story. As she researched this book, talking with seal friends and seal foes alike, she reports on some excellent advice given by an Atlantic white shark researcher, wildlife advocate, and educator: *"Just listen."* This Alix has done, using the same respect, empathy, and open mind with which she observes the seals.

As you join Alix in the pages ahead on her yearlong quest to understand seals and their complex relationships with human beings, she's going to share some poignant and revelatory stories. Some are tragic. Some are frustrating. Some are funny. Some are sweet. *Just listen.* All these stories contain truth. And we will need to listen to everyone's truth—including the seals'—if we are to successfully share our sweet, overcrowded, blue and green world together.

PREFACE

IT WAS LATE JULY of 2020, and Maine was in the midst of a brutal heat wave. Daily temperatures stretched well into the nineties, and the air was thick with humidity. To escape the sizzling heat, Mainers flocked to local beaches, plunging into the frigid salt water with uncharacteristic enthusiasm. But on the morning of July 27, the days of carefree summer swims came to an abrupt and tragic end. Julie Holowach, a sixty-three-year-old New Yorker who had long spent her summers in Maine, was swimming with her daughter just twenty yards off the coast of Bailey Island, in Harpswell, when she was killed by a great white shark. It was the first known fatal shark attack in Maine's history.

In response to the attack, state officials dispatched marine patrol boats and aircraft to scan Maine's coastal waters. Swimmers at nearby beaches were warned not to venture beyond their ankles. A wave of shock and fear reverberated up and down the coast. Few had known there were white sharks swimming in Maine's nearshore waters. Holowach's death was a terrible wake-up call.

I had been living in Portland at the time, not far from where the shark attack had occurred. I braced myself for what I assumed would come next—a *Jaws*-era anti-shark movement. Instead, to my surprise, the spotlight fell on seals. Headlines in prominent news outlets highlighted growing concerns over rising seal numbers: "Increase in Seal Population Likely Attracting More Sharks to Maine Waters"; ". . . Debate Intensifies Over Culling Seals." During a press conference, Maine's Department of Marine Resources commissioner, Patrick Keliher, said, "This is a predator-prey relationship issue. It's the presence of seals that [is] really the driver."

The truth was, I'd never thought much about seals. They seemed harmless, curious, playful—reminded me a bit of my dog. It was clear they were intelligent, often a mainstay at zoos and aquariums, trained to perform various tricks in front of throngs of tourists in exchange for fish. Occasionally, I'd see them sunbathing on rocks or ledges along the shoreline, their heads and tails raised as they lounged on their bellies, like blubbery bananas.

But seemingly overnight, spotting seals anywhere along the nearly 3,500 miles of Maine coastline no longer instilled a sense of wonder and amusement for Mainers. Instead, seals conjured our most primal fears. They served as a visible reminder of the great white predators lurking beneath the surface.

As I dug into the research, I discovered it wasn't just shark-fearing beachgoers who were frustrated by rising seal numbers. Many fishermen throughout New England were outraged over what they perceived to be out-of-control seal populations feasting on their catch and occasionally damaging their fishing gear.

On the other hand, wildlife advocates and nature lovers were thrilled to see thriving populations of seals along the shoreline. Their enthusiasm for these potato-shaped, big-eyed creatures powered a thriving tourism industry in New England—there were seal boat tours, seal walking tours, seal stuffies and post cards at gift shops, seal-shaped cookies at my local bakery (which were delicious, I should add—all part of my research).

While seals have coexisted with humans for millennia, in

recent centuries, seal populations in many regions of the world were hunted to near extinction. The U.S. was no exception. By the mid-twentieth century, gray seals had been virtually eliminated from New England waters, and the population of Atlantic harbor seals was reduced to a tiny fraction of its former size. Since then, largely thanks to the 1972 Marine Mammal Protection Act, gray and harbor seal populations in the region have made a tremendous comeback. Today, there are roughly 61,000 Atlantic harbor seals and 28,000 gray seals living along the East Coast of the U.S. But their expanding populations have fueled a rise in interactions with humans, and reignited a heated controversy I didn't even know existed.

I began to wonder, what happens if the laws and policies we enact to protect a species—to undo centuries of human exploitation—work *too* well? How many seals is the right amount, the *natural* amount? What do we know about the life cycles of seals, their behavior, their ecological role? How will our relationship with these animals shape their future, and ours?

THROUGHOUT HISTORY, OUR UNDERSTANDING of seals has been dramatically shaped by the stories we tell about them. Their survival, their destruction, is often guided by the competing interests and demands of humans.

What began as a backyard investigation into the growing populations of seals in New England evolved into an obsessive quest that took me across North America—from the seal-speckled craggy coast of Maine to a mythical, storm-ravaged island in Atlantic Canada, from human-seal battlegrounds in the Pacific Northwest to the sharky waters of Cape Cod.

This book follows my year of reporting as I explored the clashing human perceptions of this iconic marine mammal. Through the changing seasons, I followed the lives of seals, discovering as much as I could about how they interact with their environment. By learning from the people closest to seals, I tried to make sense of our relationship with these mysterious yet familiar creatures.

In the end, this is just one person's story. It is shaped and limited by my own history, experiences, biases, and gaps in knowledge. While I did my best to give voice to competing perspectives, I'm of the mindset that reporting can never be truly objective, as hard as we might try to make it so.

What follows is one of the most complex conservation tales that exists, told as I saw it playing out in real time. And it all begins, or at least I think it's appropriate to begin, with a talking seal named Hoover.

A YEAR WITH THE SEALS

PART ONE
AUTUMN

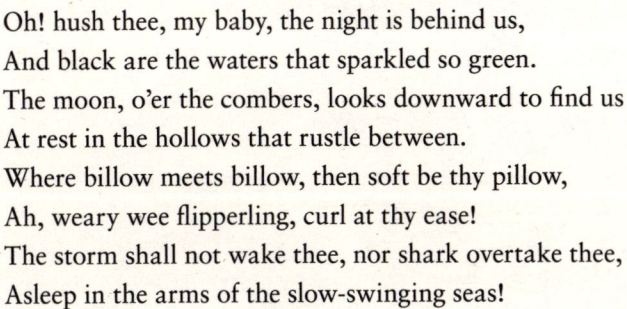

Oh! hush thee, my baby, the night is behind us,
And black are the waters that sparkled so green.
The moon, o'er the combers, looks downward to find us
At rest in the hollows that rustle between.
Where billow meets billow, then soft be thy pillow,
Ah, weary wee flipperling, curl at thy ease!
The storm shall not wake thee, nor shark overtake thee,
Asleep in the arms of the slow-swinging seas!

—RUDYARD KIPLING, "SEAL LULLABY,"
FROM "THE WHITE SEAL"

MAP OF NORTHWEST ATLANTIC

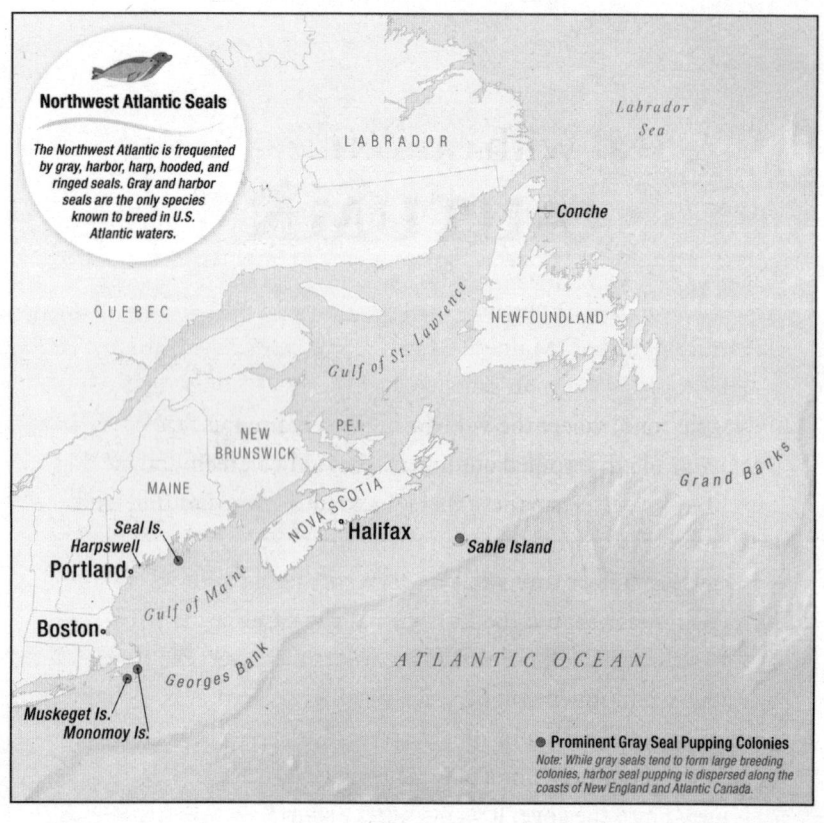

1

A Seal with Something to Say

ON A CLEAR, QUIET morning in 1971, in the remote fishing village of Cundy's Harbor, Maine, Scottie Dunning heard the unmistakable bark of his German shepherd. Dunning followed the sounds down the road, where the white pines and red maples gave way to a rocky beach that spilled out into the bay. There, he found his dog standing beside the smallest harbor seal he'd ever seen. The wide-eyed pup was no bigger than a watermelon.

Dunning wasn't sure what to do with the seal, which couldn't have been older than a day or two. He returned home to call his brother-in-law, George Swallow, who lived just down the road. Swallow was known for his love of animals and was prone to adopting an eclectic mix of injured wildlife. Together, the men searched for the seal's mother along the rockweed-covered rocks that jutted into the cove. It wasn't long before they found her—her lifeless body lay across a ledge. She had been shot.

At the time, seal shootings were nothing new, particularly in New England fishing communities. Fishermen were known to be hostile to seals, as they were viewed as competition for commercially important fish like cod and herring.

Swallow knew the seal pup wouldn't stand a chance without its mother—harbor seals rely exclusively on their mothers' high-fat milk for their first four to six weeks of life to survive—so he decided to bring him home. He and his wife, Alice, tried to feed the seal warm milk from a baby bottle, but he refused to eat until a local fisherman advised Swallow to grind up some fish in a meat grinder. The tiny seal was soon sucking up mackerel mush like a vacuum. They named him Hoover.

At first, Hoover lived in the Swallows' bathtub, but Alice soon tired of sharing their only bathroom with a fast-growing and pungent marine mammal, so she asked her husband to find a more suitable location to house the pup. Swallow moved him to the freshwater pond behind their house, built a tent-like shelter next to the pond for him, and surrounded it with wire fencing to protect him from predators at night.

Hoover adjusted quickly to his new home, playfully swimming in circles in the pond before hauling out of the water onto a flat rock to warm up in the sun. If Swallow was late in giving Hoover his breakfast, the young seal would galumph from the pond to the house, belly hopping his way up the back steps. Swallow would open the door to find Hoover staring up at him. "Hello there," he'd say to the seal, then cart Hoover back to the pond in a wheelbarrow. According to Alice, who recounted the story in her book, *Hoover the Seal, and George,* Hoover enjoyed this so much, he'd often return to the house for another ride.

Swallow enjoyed bringing Hoover along for car rides while running errands in town. Alice recalled one such trip to a hardware store in the nearby town of Brunswick. Swallow left Hoover in the car with the window open while he went into the store. "As he was paying for his merchandise a lady beside him let out a scream and pointed to the floor," wrote Alice. "There was Hoover! He had somehow gotten out of the car and come to find George."

Swallow often spoke to Hoover while spending time with him at the pond. If Hoover was hiding in the cattails, Swallow would holler for him to "get out of there, and come over here," and Hoover would rush toward his human caregiver and greet him with a wet kiss. It wasn't long before word spread throughout the neighborhood of Cundy's Harbor about the newest addition to the Swallow household. Hoover quickly became a beloved member of the community, and local children often popped by to visit him. One day, Hoover made a noise that sounded to Swallow an awful lot like "Hello, dere." Swallow gave Hoover a funny look but otherwise didn't think much of it. But days later, some of the neighborhood children told Swallow that Hoover had said hello to them. Eventually, Swallow was able to ask Hoover, "What's your name?" and he'd respond, "Hoover," although it technically sounded more like "Hoov-ah." The seal mimicked the same gruff New England accent as Swallow, who was born in a small town just outside of Boston.

As Hoover grew, so too did his appetite. At first, the Swallows were able to keep the young seal well satiated thanks to the seasonal arrival of Atlantic mackerel along the coast of Maine. In the late spring, these silvery, schooling fish gathered in large numbers in protected coves and harbors in search of food, and the neighborhood children would collect the fish in buckets to bring to their flippered friend. But by the late summer, as the fish began migrating back to the open ocean, it became a challenge to secure enough food for Hoover.

For a time, Swallow fed him frozen whiting he'd collected from a company in Portland, along with a variety of vitamins and supplements to maintain his health—a recommendation from the animal care team he'd called at the New England Aquarium. But by August, Hoover's need for up to fifteen pounds of fish a day had become too costly for the family to maintain, so Swallow called the aquarium once more to ask if they could take him in.

On the day they drove Hoover to Boston, Alice was in tears.

Years later, during an interview for a *New Yorker* article about Hoover, Swallow recalled that before they left the aquarium to return to Maine, ". . . I told a fellow there, 'I think he can talk,' but he gave me such a look I never mentioned it again."

ON A WARM EVENING in early autumn, I embarked on the short drive from my house in Midcoast Maine to Cundy's Harbor. It was my favorite time of year. For a few short weeks, after the summer visitors have left but before the leaf-peeping tourists arrive, Maine settles into a period of relative calm. The nights are cool enough for cotton quilts, the days warm enough for impromptu swims during quiet coastal hikes. But that evening, I was singularly focused on my peculiar, self-assigned fact-checking mission.

A low haze had gathered along the roadway by the time I arrived at my destination: Cranberry Horn Cemetery. I turned onto the narrow dirt road that ran between the rows of gravestones. The mist hovered like a blanket above the hollow ground, adding an ominous Stephen King vibe to the evening. Thankfully, it didn't take long to find the gravestone I was seeking, as "SWALLOW" was written in large block letters across the center of the stone. Underneath the surname was a quote: "Keep out of the hot sun, don't take any wooden nickels." And etched across the top, just as I'd read, was a detailed illustration of George Swallow with his arm wrapped around a seal. The two creatures were gazing warmly into each other's eyes, and the words "George and Hoover" were written in small letters underneath.

George Swallow died in 1997 at the age of eighty-two, more than a decade after Hoover. While I'd become captivated by the story of a talking harbor seal that grew up a short distance from my own backyard, it seemed paradoxical. How could an animal beloved enough to have his image carved onto a man's gravestone also be despised enough for someone to shoot his mother dead on a rock? What was it about seals that elicited such intense human emotion?

A FEW DAYS AFTER my graveyard visit, I drove south from Maine's historic shipbuilding city of Bath along a peninsula that stretched into the town of Phippsburg. It was midmorning, the September sun glistening off the slate blue waters of the Kennebec River. I was on my way to a seal release.

Months earlier, I'd discovered a nearby marine mammal rescue organization that responded to sick and deceased animals up and down the coast. The group, Marine Mammals of Maine, was led by a forty-seven-year-old Phippsburg native named Lynda Doughty. Lynda and her team were responsible for monitoring roughly 2,500 miles of beaches and rocky coastline—from the southernmost town of Kittery to the Midcoast town of Rockland. (Their partner organization, Allied Whale, covered the remainder of the coast, from Rockland to the Canadian border.)

Marine Mammals of Maine also operated the only marine mammal rehabilitation center in the state, with the largest capacity for seals in the Northeast. As it turned out, the rehab center was just a fifteen-minute drive from my house. I was eager to learn more about this group and their warm, friendly, and impressively hard-to-pin-down founder and executive director, Lynda. I'd been trying for weeks to set up a time to meet her and hopefully see her staff in action. So I was thrilled when she let me know that they'd be releasing two harbor seals back into the wild that very morning. And I was invited.

The populations of gray and harbor seals—New England's two resident seal species, with harp and hooded seals as occasional visitors—were thriving along the northeast coast of the U.S. I'd recently spoken to a veterinarian and marine wildlife expert in Massachusetts who had been involved in seal rehab efforts in the past. When I asked him why his team no longer focused on seals, he said that one of the big reasons was the lack of available funding to support the work. Given how well seal populations are faring, he told me, they aren't priority species for limited state and federal grant opportunities, and the work had become hard to justify. It would be like rehabbing deer or squirrels, he explained. Instead, his team had shifted their focus to supporting marine species whose populations were threatened or endangered.

I could see how that made sense. And yet, Lynda and her team dedicated considerable time and effort to seal rescue and rehab efforts, and they weren't the only group invested in that work. I was interested in her perspective.

I arrived at Head Beach, on the southwest tip of the Phippsburg peninsula, just after nine a.m., surprised to discover there were already more than two dozen cars and trucks parked in a row. A small crowd had gathered at the entrance to a beach path that led over the dunes, while others had made their way to the water's edge, anxiously awaiting the arrival of the seals.

It wasn't long before the Marine Mammals of Maine truck—a gray Ford crew cab with a shell over the bed, the perfect human-seal transport vehicle—appeared in the distance. As it approached, I spotted Lynda's bright blonde hair behind the wheel. She backed the truck up to the entrance and several staffers quickly hopped out, arming themselves with large plastic boards that could be used, if needed, to gently guide seals toward the surf. Lynda, meanwhile, began urging the enthusiastic crowd to move to the beach to help clear the path for the seals. She then lowered the tailgate of the truck, revealing two large, sand-colored kennels with metal cage doors. For a moment, the kennels appeared empty, their contents dark, with no perceptible sounds or movements. But the darkness soon began to take shape, as two sets of large, glistening eyes peered out from behind the cage doors.

Volunteers carried the kennels over the dunes to the beach, setting them a few feet apart, a short distance from the surf. Lynda then turned to face the crowd. "I just want to say a few words," she said. Unfortunately, her words were largely muffled by the roar of crashing waves and wind whipping across the dunes, but I did make out the seals' names: Number 64 and Number 87. I later learned these numbers reflected the order in which they were rescued—they were the sixty-fourth and eighty-seventh seals the team had responded to that year. After undergoing several months of rehab support, the now-healthy seals were ready to return home.

Lynda cupped her hands around her mouth and shouted over the wind, "Is everyone ready for the countdown? Five . . . four . . . three . . ." As we reached zero, the staff unlatched the kennel doors. A few seconds later, two shiny, round heads cautiously poked out. Before long, the seals were galumphing toward the surf like determined caterpillars. The dynamic duo splashed through the shallows, using their front flippers to steer their rotund bodies toward the depths. But as they approached an oncoming wave, they hesitated, then turned and quickly retreated. They swam horizontally along the edge of the surf, searching for a calm portal to the deep blue beyond the breakers.

As the crowd watched the seals adjust to their new habitat, I approached two young women standing nearby to ask how they'd heard about the release. As it turned out, they had both been interns at the rehab center the preceding summer and had taken time off from their fall semesters to watch the seals they'd spent months caring for finally be released back into the wild. One of the women, Ciara Dunn, a student at Trinity College, said that Number 64 had kept them on their toes with her antics that summer. She was notorious for diving to the bottom of the tank to evade them when it was time to administer her medicine.

But the story they shared about Number 87 was particularly

alarming. The seal had been born on a crowded beach in Southern Maine, and had been harassed by overly attentive humans who were enamored with the tiny, big-eyed pup. One beachgoer became concerned about the seal's welfare and called the Marine Mammals of Maine hotline to alert them. The staff quickly dispatched local volunteers to stand watch until they could get there to assess the seal's condition. For a time, the volunteers successfully discouraged people from approaching the seal, but during a volunteer shift change, the pup was briefly left alone. During this window, Number 87 was carried to the surf as beachgoers attempted to put her "back" in the ocean. In their early days and weeks of life, harbor seal pups require long stretches of rest time out of the water, and their mothers often leave them alone while they go off to forage. By the time the staff arrived, the pup was in such a state of stress and exhaustion, they had no choice but to admit her to the center.

As seals and humans have expanded along Maine's coastline, human-seal interactions like these have been on the rise. Without the support of Marine Mammals of Maine, it was unlikely either of these animals would have survived.

The two seals eventually dove beneath the waves, then popped their heads up to look back at the humans quietly cheering them on from the beach. We glimpsed them several more times as they swam farther out to sea, until we finally lost sight of them, well beyond the breakers.

Driving home after the release, I thought about the crowd's collective excitement to see the two seals return to their wild waters. I had felt it, too: the anticipation, the enthusiasm, the sense of connection to a creature that seemed at once familiar and foreign.

THE CLOWNS OF THE marine mammal world, the personalities of the sea, seals have endeared themselves to humans for millennia.

Throughout history, we've been fascinated by their inquisitive nature, intelligence, and silly mannerisms. Humans are hardwired to care more about animals that resemble us, said Jon Mooallem, author of *Wild Ones*, a phenomenon known as "phylogenetic relatedness." As Mooallem explains, this may be because ". . . we assume that a creature that looks vaguely like us will have similarly high capacities for thought, pain, and feeling."

As it turns out, "charismatic appeal" is a quantifiable metric in conservation science—one of a host of measures scientists and policymakers use to rank which species' protection we should devote limited resources to. In a world where wildlife extinctions are rampant, the emotional connection humans have with an animal can affect its very survival. As evolutionary biologist Stephen Jay Gould said, "We are, in short, fooled by an evolved response to our own babies and we transfer our reaction to the same set of features in other animals." Seals' round faces, large eyes, floppy movements, and plump bodies have, in some ways, worked to their advantage when it comes to eliciting human compassion.

Seals are part of the order of mammals known as *Pinnipedia*, Latin for "fin- or feather-footed"—a group that also includes sea lions and walruses. It's believed that the ancestors of pinnipeds were bear- and weasel-like landlubbers that were increasingly drawn to the water for its abundance of food. Over time, they adapted to a life aquatic. But pinnipeds have maintained close ties to the terrestrial world, as they spend long stretches of time on land. As a result, we feel we know them better than other marine mammals, which, as naturalist and conservationist Farley Mowat points out in his book, *Sea of Slaughter*, is partly evidenced by the names we've given them: sea dogs, sea lions, and sea elephants, to name a few.

Gray and harbor seals are among eighteen seal species that comprise the family of pinnipeds known as *Phocidae*, also referred to as "true" or "earless" seals. The other two families are *Otariidae*, or "eared" seals, so named for their possession of external ear flaps (true seals have flat slits for ears), which includes fourteen species of sea lions and fur seals, and *Odobenidae*, which includes just one species—the walrus.

Eared seals have large front flippers and hind flippers they can turn to face forward, allowing them to move with relative ease on land. True seals, on the other hand, have short, stubby front flippers, and their hind flippers only face backward. If you picture a seal at an aquarium clapping its flippers and "walking" alongside its tank or pool, that's likely a sea lion or a fur seal. Asking a true seal to clap its front flippers would be like asking a *Tyrannosaurus rex* to give you a bear hug. On land, true seals move like giant, blubber-bellied caterpillars, shifting their weight from back to front as they shuffle forward. They use their front flippers mostly for balance and a little extra oomph. Despite these comically clumsy belly shuffles, however, seals are surprisingly swift on land and have been known to outrun humans.

When it comes to vocalizations, sea lions take the cake as being among the most vocal mammals in the world, with their distinct "ark ark" barks. Although less chatty than sea lions, the northern elephant seal is one of the loudest mammals ever recorded. Elephant seals get their name not from their massive size but from their noses, which resemble an elephant's trunk. When male elephant seals reach sexual maturity, they develop a proboscis that helps to attract females during mating season, in part by enabling them to produce ear-piercing, sexy roars.

One reason for pinnipeds' unique ability to vocalize is to help mothers find their pups in crowded breeding colonies. Some pinniped mothers and pups recognize one another's distinct calls long after the pups have grown up and separated. Scientists in Alaska found that northern fur seal mothers can identify their pups' vocalizations even after four years apart.

AFTER HOOVER'S ARRIVAL AT the New England Aquarium in 1971, aquarium staffers' skepticism about the seal's ability to talk appeared, at first, to be justified. Among the thirty-three species of pinnipeds, harbor seals are known as the "quiet seals," and Hoover was true to his species' reputation. But after a couple of years, he began emitting a string of strange and, for the aquarium staff at least, startling vocalizations, including growls, squawks, shrieks, and "blood-curdling screams," according to one record. Then, in mid-November of 1978, a staff member recorded the following observation: "Says 'HOOVER' in plain English. I have witnesses."

Before long, Hoover was entertaining throngs of tourists with his vocal talents. Bobbing vertically like a soda bottle in the pool he shared with several other harbor seals, Hoover would tilt his head backward until his eyes were submerged, and in a deep, rumbling voice that appeared to come from the back of his throat, his mouth hardly moving, say, "Come ov-ah hee-yah" or "Hello dere" in his signature accent. He would speak only of his own accord, when he felt the urge, which happened more often during mating season. And it wasn't uncommon for Hoover to wait until an unsuspecting visitor strolled past the seal exhibit before abruptly calling out, "Get outta there!" in a deep, raspy voice, followed by a humanlike belly laugh.

According to Alice, ". . . if you ever heard Hoover speak he sounded just like George." Andrew Hiss, the *New Yorker* journalist who penned the 1983 Hoover article, agreed: "Hoover sounds *exactly* like Mr. Swallow, except, as Mr. Swallow himself points out, when Hoover says 'Hoover.' Then, Mr. Swallow says,

Hoover sounds more like the daughter of a neighbor of his who used to hang around the pond and liked to tell Hoover what his name was."

In the summer of 1981, George and Alice Swallow visited Hoover at the aquarium—more than a decade after they'd adopted him as a pup. Swallow positioned himself on the side of Hoover's pool and yelled, "Hey, stupid!" Upon hearing the familiar greeting, Hoover was so excited that he quickly swam over, took Swallow's hand in his mouth, and attempted to pull him into the pool.

Hoover died in 1985 at the age of fourteen. His death, on the younger side for a captive harbor seal, was said to have resulted from complications related to his annual molt, when seals shed their fur to make way for a new coat.

The day after Hoover died, the *Boston Globe* published his obituary with the headline, "Hoover Will Talk No More." While he had sired six pups during his fourteen-year tenure at the New England Aquarium—Trumpet, Amelia, Joey, Lucifer, Cinder, and Spark—none of his offspring exhibited a desire nor an ability to mimic human speech.

IN 1971, THE YEAR Swallow took Hoover in, the population of harbor seals off the U.S. Atlantic Coast had been reduced to a tiny fraction of its former size, and gray seals had been virtually eliminated from U.S. waters. But in the decades since, the populations of these two species have dramatically increased. According to the 2023 stock assessment reports from the National Oceanic and Atmospheric Administration, or NOAA, the breeding population of gray seals in the U.S. is roughly 28,000, most of which pup and mate in large colonies on beaches in Nantucket and Cape Cod, and on remote islands in Maine. The population of harbor seals in U.S. Atlantic waters, meanwhile, is an estimated 61,000, the vast majority of which give birth along Maine's rocky coastline.

This dramatic increase in seal numbers has been largely attributed to the passing of the Marine Mammal Protection Act in 1972—just a year after Hoover's mother was killed. The act, which bans the hunting, harassment, capturing, or killing of any marine mammal in U.S. waters, or by a U.S. citizen on the high seas, was groundbreaking legislation at the time, paving the way not just for the recovery of gray and harbor seals, but for many other species, including gray whales, humpback whales, northern elephant seals, and fur seals.

Anna Magera, a researcher at Dalhousie University, and her colleagues analyzed trends for ninety-two marine mammal populations. According to their findings, published in 2013 in the scientific journal *PLOS One*, 42 percent were shown to be increasing, 10 percent decreasing, with the remainder either unknown or showing no change.

In a world where wildlife populations have been disappearing at an alarming rate, stories about recovering species are a relatively new phenomenon. This is particularly true when it comes to marine wildlife, given our long history of exploiting the seas through whaling, the fur trade, and industrialized overfishing. Thriving seal populations have been hailed as a victory by many conservationists and wildlife advocates, the result of one of the most successful pieces of environmental legislation in history, but the reality has been more complicated. As seal populations have rebounded, so too have conflicts with humans—particularly fishermen.

Fishing communities throughout New England have struggled with increasing seal numbers, as seals have demonstrated a healthy appetite for many of the same fish that fishermen are targeting. They won't turn down the opportunity for a quick and easy meal, even if it requires chomping through a fishing net to gorge on an all-you-can-eat sushi buffet. And conflicts between seals and fishermen are by no means limited to New England. Similar challenges are playing out in coastal communities around the world as seals and humans attempt to coexist on either side of productive fishing grounds.

Seals live on every continent on Earth, although most prefer colder waters. Some species, such as Hawaiian monk seals and Baikal seals (the latter being the only species that lives exclusively in fresh water) have relatively small ranges. But other species, including gray and harbor seals, are more widespread. Gray seals live in coastal areas on either side of the North Atlantic, while harbor seals have the widest distribution of any seal species, inhabiting coastal regions across much of the Northern hemisphere.

I spoke to Rob Lambert, a British environmental historian at the University of Nottingham, who told me that in the U.K., gray and harbor seals are the most deeply contested of all wildlife species, given how polarized stakeholder viewpoints are. I had called Rob to get his perspective on what I'd begun to think of as "the pinniped paradox"—the ways in which seals inspire devotion and hostility in seemingly equal measure. "Ultimately, the story of seals is less about nature than it is about human nature," he said.

Rob told me about a Cornish fisherman he'd met who used to shoot gray seals as target practice for fun in his youth. The fisherman and his friends had a point system—more points were given for a seal pup, as opposed to an adult, but the most points were reserved for seals with rehab markings. From their perspective, Rob explained, the very concept of rehabbing seals was a personal affront. The idea that these animals were being rescued, cared for, filled with milk, snuggled up beside seal stuffies, and sent on their way in perfect health was too much for some fishermen to bear.

The fascinating thing about Rob's story was that the fisherman who shared it with him did so at the end of a seal-watching nature tour—one Rob had been leading. The fisherman, who was in his seventies at the time, had joined the tour with his seal-adoring granddaughter. He told Rob that he was horrified by what he had done in his youth.

When it comes to human-seal conflicts, Rob explains the pinniped paradox in simple terms: "Those of us who take great delight in seals do so without any economic loss." I thought about the seal release I'd watched in Phippsburg, the dozens of delighted seal fans cheering them on as they galumphed into the surf. Days before the event, I'd called my sister, who lives with her husband, a commercial fisherman, and their three-year-old son in Gloucester, Massachusetts. "You should come up to Maine to see the release," I'd said, knowing how much my nephew would love it. I then asked if she thought my brother-in-law, a huge fan of animals and nature, might want to join. In response, she burst into laughter. "I actually can't imagine anything he'd hate more than a seal release," she said.

THERE SEEMED TO BE so many competing perspectives on seals that it was hard to make sense of how, or if, they fit together. Everyone I spoke to had such strong feelings about them, but generally that had to do with the very local, specific ways they interacted with seals. Was it true that seals were destroying inshore fisheries? Had their populations reached unsustainable levels? Were seals luring sharks to busy New England beaches?

I wanted to find out what marine mammal biologists thought about rising seal numbers—and what fisheries and shark scientists thought, for that matter. It was hard to imagine wildlife officials responding to calls for a seal cull—that it would be possible for the species that helped to inspire the Marine Mammal Protection Act in the first place to now trigger its dismantling.

Yet at the same time, the political landscape has dramatically changed since the 1970s environmental movement. Support for environmental policy has become an increasingly divisive issue, and some of the protections put in place over fifty years ago have been significantly weakened or eliminated in recent years, often fueled by corporate interests. It wasn't inconceivable to think that the days of blanket marine mammal protections could be numbered.

Politics aside, I wanted to understand our relationship with these iconic sea creatures, in part because I'd started to develop my own fascination with seals—their curiosity, their goofy behaviors. But I also had a lingering sense that seals had something to teach us about the broader ecological challenges we're facing today.

One thing was for sure—seals had their fair share of fanatics and foes. Even Hoover wasn't immune to controversy. Although in his case, his nemesis wasn't human. Rather, it was a fellow harbor seal named Andre.

2

The Traveling Seal

THE CHALLENGE WITH OBSESSIVELY researching all things seals was that the more I learned, the less fun I became in human social circles. When a friend jokingly imitated a seal by barking and clapping her arms like flippers at a dinner event, instead of politely laughing, I spent the next ten minutes explaining the biological differences between true seals and eared seals, emphasizing the ways in which she had mistakenly imitated a sea lion, rather than a seal. By the time I'd reached my closing arguments, which consisted of a personal rendition of seal vocalizations complete with howls, hisses, and frog-like croaks, the room had gone silent—all eyes were on the crazy seal lady.

But it was also true that I'd amassed a collection of stories about charismatic seals by this point, and unlike my seal mimicry attempts, these never failed to entertain. It always seemed the most popular stories were the ones that featured seals befriending or behaving like humans. The more seals seemed like us, the more enamored we became.

YOU'D BE HARD PRESSED to find a more famous pinniped in all of New England than the ball-bouncing, hoop-leaping, television-watching, high-fiving harbor seal known as Andre. But Andre's allure stemmed from more than his behavioral talents. One of the most captivating things about this seal was his relationship with the human who adopted him—a salty, eccentric Mainer named Harry Goodridge.

In late autumn each year, beginning in the early 1970s, as the last of Maine's migrant birds flew south for warmer climates, Goodridge would load Andre into the back seat of his station wagon in the small coastal fishing village of Rockport, and the two creatures would begin their own southward migration by way of Interstate 95. As the wild world braced for winter, man and seal made their way to the New England Aquarium.

BEFORE ANDRE CAME ALONG, Goodridge had been searching for years for the perfect fin-footed diving companion. An arborist by trade, Goodridge was Rockport's harbormaster and an avid scuba diver. By the late 1950s, he had become, by his own admission, obsessed with the notion of befriending a harbor seal. "I couldn't quite put down the curious feeling that a harbor seal had something to tell me," he wrote in his autobiography, *A Seal Called Andre*, which he coauthored with his friend, local writer Lew Dietz. "I was completely hooked on harbor seals, and over the long winter I came to accept my addiction."

Unfortunately, Goodridge's obsession resulted in unintended, but nevertheless fatal, consequences for several seals, as he learned about the challenges of properly feeding and caring for what were, of course, wild animals, despite their seeming familiarity to humans.

Goodridge first encountered Andre in 1961 while out on his boat in Penobscot Bay. After scanning the water in search of the pup's mother, he scooped up the days-old harbor seal in a net and brought him home. While Goodridge's wife, Thalice, and their five children had grown quite accustomed to his eccentric animal adoptions—bats, seagulls, flying squirrels, and skunks, to name a few—Andre was something special.

Goodridge built a large floating pen for Andre to keep him away from boats and other activity in the busy harbor. The pen included a platform where Andre could "haul out" of the water, a natural behavior where seals come ashore, sometimes just to rest or to regulate their body temperature.

In the evenings, Goodridge would visit Andre in the harbor, standing on the platform as he taught him to perform various tricks in exchange for fish. Before long, Andre had learned to wave, blow raspberries, bounce a ball off his nose, and cover his face with his flipper to feign embarrassment. Eventually, word spread about these evening training sessions, and locals began to show up in increasing numbers to watch the duo perform.

After the busy summer season, as things began to quiet down in the harbor, Goodridge would release Andre from his floating pen, letting the seal free to come and go as he pleased. But as the years went by, Andre began to draw the ire of local fishermen with his shenanigans. He'd grab at fishermen's oars as they tried to row their dinghies to the moorings that held their boats. He'd pull off divers' flippers and jump into boats. Frustrations grew and Goodridge began to fear for Andre's safety.

Goodridge called the New England Aquarium to see if they'd consider taking Andre in for the winter to keep him out of trouble. Andre could then return to Rockport in the spring. They agreed.

The charismatic seal adapted quickly to his new digs. Aquarium visitors were delighted by Andre's impressive repertoire of tricks, and he was a hit with the other harbor seals in the exhibit, with just one exception: Hoover. When Andre first arrived at the aquarium in 1973, Hoover had been the dominant male and wanted nothing to do with the popular new entertainer. Hoover became so depressed, in fact, that the trainers were forced to temporarily remove him from the exhibit to help ease the transition. (A few years after Andre arrived, however, Hoover made his way back to center stage by showcasing his own unique talent: the uncanny ability to mimic human speech.)

In the spring, when it was time to transport Andre back to Rockport, Goodridge surprised the aquarium staff when he announced that Andre would be swimming back on his own. The staff cautioned Goodridge against the idea. The idea of a harbor seal swimming hundreds of miles from Massachusetts to Maine and somehow finding his way back to Rockport's tiny harbor seemed unlikely at best. While little was known about harbor seals' navigational abilities at the time, Goodridge was adamant that Andre could do it, if he wanted. He'd accepted that Andre might choose to continue life as a wild seal.

Andre was released at a nearby beach in Massachusetts. Within days, Goodridge's phone began to ring with reports of Andre sightings as the seal made his way up the coast, past the sandy beaches of Massachusetts and New Hampshire to the rocky coast of Maine and, eventually, all the way up to Rockport.

Some attributed Andre's return to his guaranteed food source, but others, including Andre's primary trainer at the aquarium, Annie Potts, felt there was more to it. "I think that there was a bond between Harry and Andre," Potts said in the PBS documentary *The Seal Who Came Home*. "I think Andre went home for that much more than he went home for a herring sandwich."

The truth was that scientists had no way of knowing. It was clear Andre didn't *need* Goodridge—the seal had long since learned to hunt for fish on his own. But for the remainder of his life, every fall, Goodridge would deliver Andre to the aquarium, and every spring, the half-migratory harbor seal would swim hundreds of miles "home."

Andre seemed to be caught somewhere between the lives of

a captive and wild seal. Perhaps his devotion to Goodridge and his annual returns to Rockport were a case of Stockholm syndrome, or perhaps, as many believed, they were evidence of a deep human-seal bond.

Either way, Andre's story was beloved, not just in Maine but across the U.S. and around the world. Students in Wisconsin created an Andre the Seal fan club, newspapers in the Bahamas published articles reporting on Andre's journeys, and his story was featured in a Japanese documentary. Andre even inspired a Hollywood film released in 1994 by Paramount Pictures, although to the frustration of his devoted fans, the role of Andre was played by a sea lion.

OUR FASCINATION WITH WATCHING and learning about other species is connected to the very notion of what makes us human. According to evolutionary biologist David Barash, throughout our evolutionary and recent past, our survival and well-being have depended on our relationships with other animals. We are "attuned to the presence as well as the habits of other beasts, especially large and dangerous ones," he wrote in a 2014 article in *Aeon*. Our ancestors were driven by a desire for self-preservation, and it was critical to keep a close eye on potential predators. But there was also value in closely observing animals that might serve as prey. By watching their movements and behaviors, early humans could learn the most effective ways to approach them. Beyond the predator-and-prey dynamics, we relied on domesticating animals to provide food like eggs and milk, or for assistance in hunting. We even relied on them for warmth, ". . . not only via their skins and fur," wrote Barash, "but also their literal bodies, cuddling closely with our Pleistocene ancestors during those long, challenging Ice Age nights." As humans, we possess an innate desire to know, watch, study, and yes, even befriend animals.

Over time, our relationship with animals shifted, but our fascination with them continued. Increasing urbanization created distance between humans and wildlife, and people were no longer directly interacting with wild animals in their natural habitat. Perceptions of animals among urban dwellers began to shift from seeing them for their utilitarian roles—predator, prey, domesticated livestock—into seeing them as emotional beings that symbolize the sanctity of the natural world. Today, even the meat we buy at the grocery store bears little resemblance to the animals from which it came—the names alone (beef, pork, sausage) distance buyers from the notion that we're consuming the refrigerated flesh of other beings.

And yet as much as we've separated ourselves from wild animals, we've become increasingly attached to our pets. Dogs, cats, fish, ferrets, donkeys, emotional support peacocks—many of us treat our pets like full-fledged members of the family. We dole out enormous sums of money for their food and medical care. We shower them with affection as well as toys, blankets, therapeutic cooling pillows—sometimes our own—to sleep on at night. I'm still shocked by the notion that I'm unable to claim my beloved dogs and cat as dependents during tax season.

People who own pets are thought to be more caring and empathetic than their counterparts, a friend once told me. She was referencing the increasing numbers of photos of men with pets featured on dating apps, but her theory is backed by data—studies on heterosexual dating app preferences indicate that men are perceived to be more attractive when they appear with dogs in their pictures. This has even prompted some men to borrow dogs for photos to increase their likes and matches, a phenomenon known as "dogfishing" (can't make this up).

Seeing and interacting with animals have well-documented mental health benefits, and in some cases, people who have trouble connecting with humans can find happiness and fulfillment in their relationships with animals—a concept that may have resonated with Harry Goodridge. "I've been called a loner and maybe I am," Goodridge wrote in his book about life with Andre. "I have no objection to people, but I must admit I relate better to animals. Perhaps it's simply that animals act sensibly."

While we might be obsessed with our pets, we're similarly fascinated by watching animals in their natural habitat. Whether it's through the soothing voice of Sir David Attenborough narrating the actions of the planet's "most extraordinary creatures" or the gripping accounts of Jane Goodall and Dian Fossey, whose research changed how we think about primates, we're captivated by the lives of wild animals. As ecologist Paul Shepherd wrote, "The human species emerged enacting, dreaming and thinking animals and cannot be fully itself without them."

SEALS OFFER US A sense of both the exotic and the familiar. While their underwater lives are largely a mystery, they look and act like aquatic versions of our domesticated dogs (hence the term sea dogs). We're drawn to seals' curious and playful personas, but more than anything, we're drawn to their impossibly huge, dark, unblinking eyes. According to Hal Herzog, a research psychologist who has spent decades studying human-animal relationships, one of the biggest factors that influences how much money people consider donating to help an endangered species is the size of the animal's eyes. Seals' eyes are adapted to quickly adjust between the dim light of the deep ocean and the bright sun at its surface. Underwater, their pupils dilate into wide circles, reflecting and amplifying weak light, which enables them to be effective hunters even in the darkest depths. Back on land, the irises close the pupils to small pinpoints, allowing them to see clearly through their round lenses.

The combination of seals' big eyes and their tendency to stare at us for long stretches has, in some ways, helped them to coexist with humans. We interpret those stares as thoughtful curiosity, perhaps even fascination, and they elicit an emotional connection within us.

IF ANYONE UNDERSTANDS THE complexities and pitfalls of our fascination with seals, it's Lynda Doughty. Not long after the seal release in September, I visited Lynda and her team at the Marine Mammals of Maine rehab center in Brunswick. I arrived midmorning and knocked at the door of the large upstairs office, where I was greeted by Dominique Walk, the group's assistant stranding coordinator, along with Dominique's two enthusiastic blue heelers, Stubby and Fin. I bent down to greet the dogs as I introduced myself to Dominique and another staffer, stranding assistant Katie Gilbert. Katie was seated at one of the five desks positioned against a corner of the large, one-room office space.

Lynda walked in moments later, and Stubby and Fin rushed to the front door to welcome her. She squealed with delight at her canine greeters, gathering Stubby in her arms and burying her face in his belly while Fin ran off to collect her Frisbee. She quickly scampered back and gently placed the Frisbee on Lynda's foot. In a single, fluid motion, Lynda shifted Stubby under one arm, picked up the Frisbee, and sailed it across the room. Fin chased after it at high speed, her nails tapping across the wooden floorboards. Dominique looked at me and shook her head. "Every. Time."

Lynda laughed as she ushered us toward a corner of the office where three well-worn couches had been arranged in a U-shape. Still holding Stubby like an infant, she collapsed onto the couch across from me, kicked off her shoes, and tucked her feet up. "So how are *you*?"

Autumn seemed to be the team's only season of relative quiet, at least when it came to seal rehab. After Numbers 64 and 87 were released in September, the dozen or so tanks, pools, and enclosures in the team's animal care center—a warehouse-sized garage space adjacent to their office—remained empty. Lynda and her six full- and part-time staff had temporarily shifted their focus from responding to seal strandings—which most often involve reports of seals that appear sick, malnourished, or otherwise in need of medical care, along with reports of dead seals—to raising money to support their operations.

With zero state funding, and federal funding only coming from a competitive grant the organization applies for each year, at most about a quarter of their budget derives from public sources. As a result, they have to raise the majority of their operating funds from individuals, companies, and private foundations. The costs add up quickly. Between food, medication, transportation, facility management, staff time, and more, rehabbing a single newborn seal pup—a process that usually takes about three to four months—can run upward of ten thousand dollars.

For Lynda, the biggest challenges seals face in Maine when it comes to humans isn't frustrated fishermen, but overly attentive beachgoers. Of the 61,000 harbor seals living along the East Coast of the U.S., more than 90 percent gather on the coast of Maine in the spring to pup and mate, which means a lot of newborn harbor seal pups in the spring and summer months. The timing overlaps with increasing numbers of tourists and summer residents frequenting the coast. "They're cute," said Lynda. "So that gets them into trouble."

Unlike other seals species, harbor seal mothers frequently leave pups alone on beaches or ledges while they go off to hunt for fish. But this natural behavior often gets misinterpreted as abandonment, and beachgoers frequently try to "help" seal pups they encounter—by picking them up and putting them in the water, covering them with a towel or seaweed, or even bringing them home. All of these actions cause high levels of stress, which can result in an animal's death. Even less invasive forms of harassment like standing too close to a seal on the beach or approaching it in a boat or kayak have proven to be just as harmful, Lynda told me. They prevent the animal from being able to safely rest, which is critical for its survival.

More than half of the seals that Marine Mammals of Maine admits to its center have suffered from some form of human interaction. "People often assume 'human interaction' means entanglement, or gunshots, or some negative experience with fisheries," said Lynda. "But it's that one selfie, or that one boat that gets too close and flushes the seals into the water." Of course, it's never just one selfie or one boat, she said. After that one boat leaves, another shows up and does the exact same thing. "If our behavior disturbs their behavior, that's harassment." And harassment equates to a federal offense, according to the terms of the Marine Mammal Protection Act. But as I'd later discover, enforcing that law is another story.

As Stubby squirmed out of Lynda's arms to chase after Fin, I looked up at the wall behind her, where dozens of framed photos of seals hung in perfectly arranged rows. "That's our new Hall of Famer wall," said Lynda, smiling. Each photo represented a seal that had been successfully rehabbed and released into the wild. Most of the photos were of gray and harbor seals, but there were a few harp seals as well.

While the collection of framed photos highlighted the very real impact Marine Mammals of Maine has had on individual animals, it represented just a tiny fraction of their work. Most of the animals in their care don't survive. During pupping season, Lynda has had to make tough choices about which seals to admit to the center, as there's only so much space for the animals. She tries to prioritize cases where animals are being harassed by humans.

She mentioned one case from years ago in Old Orchard Beach, a particularly busy beach in southern Maine, where hundreds of people had crowded around a single harbor seal pup. The rehab center had been full at the time, so Marine Mammals of Maine volunteers stood watch over the seal, pleading with beachgoers to keep their distance. But its mother never returned. Days later, after Lynda was forced to euthanize a seal at the center whose health had deteriorated, they were finally able to admit the pup. By then, the pup was so malnourished and dehydrated that it was on the brink of death. Lynda and Dominique provided round-the-clock care for months, and were eventually able to nurse the seal back to health. Its photo was now among those framed on the wall.

ANIMAL WELFARE CONCERNS STEMMING from our magnetic draw to seals and other pinnipeds isn't just an issue in Maine. Wildlife officials around the world face uphill battles when it comes to managing the public's overwhelming enthusiasm for pinnipeds.

A recent example was a 1,300-pound walrus named Freya—aptly named after the Norse goddess of love and beauty. Freya began making appearances along the coasts of the U.K., the Netherlands, and Denmark before arriving in Norway, where she decided to extend her European holiday.

By the summer of 2022, Freya had become an international darling, in part because of her tendency to welcome herself aboard (and occasionally sink) small boats docked at busy marinas in Oslo's fjord region. Photos and videos of Freya sunbathing on these vessels went viral, and it wasn't long before crowds of visitors began arriving to meet her, crowding around her for selfies, and throwing things at her to get her attention. Some even jumped in the water to swim with the walrus. Norwegian authorities cautioned the public to keep their distance from Freya, concerned her growing ease with humans had become a liability. But few heeded their warnings.

In mid-August, Freya's fan club received the shocking news that wildlife authorities had euthanized the walrus—shooting Freya in the dark of night while she slept in a boat in the marina. The government's seemingly abrupt decision sparked international outrage. But the officials, with support from the Norwegian prime minister, defended their decision. Amidst the constant human attention, they said, Freya had begun to change her behavior. She seemed to be seeking humans out. In one case, police were forced to evacuate a swimming area after Freya chased a woman into the water. Still, many of Freya's supporters felt the government should have issued stronger penalties for humans harassing the walrus. They believed officials had been far too quick to pull the trigger.

Espen Fjeld, a biologist and retired advisor for the Norwegian Nature Inspectorate, agreed with the government's decision to euthanize Freya given the increasing danger she posed to humans, particularly children who had been seen swimming quite close to the walrus. But he was frustrated that the public's obsession with a single animal distracted from more important questions around the health of the broader walrus population in the Arctic. Just months before Freya was euthanized, Fjeld told NBC News, the Norwegian government had taken steps to expand oil and gas drilling in the Barents Sea—a decision that would threaten the entire population of Atlantic walruses. Yet it had received almost no media attention, whereas Freya had become a global celebrity.

A similar story has been playing out in Australia featuring another charismatic celebrity pinniped, a juvenile elephant seal named Neil. Southern elephant seals are the largest living pinnipeds, with males growing up to twenty feet long and weighing up to nine thousand pounds—about twice the size of a full-grown walrus. They're also one of the deepest-diving animals on earth, chasing after fish and squid up to three thousand feet below the surface. While they spend most of the year foraging in the open ocean, they haul out on land beginning in September to pup and mate.

Neil was born in October of 2020 in Tasmania. By January 2024, at the age of three, he was estimated to weigh more than 1,300 pounds—roughly the same size as Freya. While male elephant seals don't reach sexual maturity until the age of six, and generally don't compete with other bulls until the age of ten, Neil has spent much of his time on land preparing for battle. His primary nemeses appear to be orange traffic cones and wildlife officers. He spends his downtime napping in the middle of busy roadways or sprawling out on the lawns of residents. Videos of Neil's antics posted on TikTok have received more than 50 million views.

Elephant seals are a threatened species in Australia, and the Marine Conservation Program has repeatedly told people to stay at least sixty-five feet away from Neil, and to keep their pets back as well. Yet there have been several reported incidents of harassment, including cases where people have poked or prodded Neil. Officials have been forced to relocate the massive seal on multiple occasions, and attempt to keep the animal's location private for as long as possible so as to discourage the public from seeking him out. The agency is eager to avoid a situation like Freya.

Back in New England, a male gray seal made headlines in the fall of 2022 when he appeared in Shoe Pond in Beverly, Massachusetts, in the middle of a busy office park. It was an unusual location, but the pond connects to Beverly Harbor through a 750-foot cement tunnel. It was likely this particular seal had followed fish through the tunnel and into the pond. Crowds of residents began showing up at the pond to visit the seal, which was affectionately nicknamed "Shoebert." Wildlife officials hoped Shoebert would leave on his own, and knew that the longer he stayed, the more likely he was to exit the pond and make his way toward the busy road. But the seal appeared to have little desire to return to the ocean.

Seal rescue teams from three states were called in to support a capture attempt. Shoebert managed to evade the trained professionals, but eventually decided to turn himself in. On September 23, at two thirty in the morning, the Beverly Police Department received a call from a security team that Shoebert was on the lawn outside their office. According to a statement from the precinct, "Shoebert appeared to be in good health and was a little sassy in the early morning hours." The seal was soon captured and brought to a nearby seal rehab center in Mystic, Connecticut, before he was released with a clean bill of health in Rhode Island. But his short time in Beverly had resulted in a sizable fan club, including the Beverly Police Department. It wasn't long before officers were wearing a Beverly Police Shoebert patch.

WHILE WE'VE GROWN SOMEWHAT accustomed to the presence of seals in New England waters, back in the 1960s, these animals were largely absent from the region, after extensive hunting had decimated their populations. But the relationship between Goodridge and Andre seemed to have renewed the public's fascination with seals. As news spread about the traveling, performing harbor seal in Rockport, people began arriving from across the country to see Andre. They bought seal T-shirts, figurines, postcards, and other Andre-themed gifts at local shops, and spent their evenings at the harbor watching Goodridge and Andre perform for the crowd. He had become a Maine celebrity.

In 1979, Andre was named "Townsperson of the Year" by the Camden-Rockport Chamber of Commerce. That same year, Maine's governor, Joseph Brennan, made the mistake of telling a room full of reporters that Andre was receiving too much publicity, urging them to focus on more official state issues. The public outrage in response to his remarks was so severe that in 1982, during Brennan's reelection campaign, he arranged a special campaign appearance in Rockport, where he issued a formal apology to Andre. Luckily for Brennan, Andre wasn't one to hold a grudge.

After Andre died in 1986, Goodridge and his family buried the seal behind their home in Rockport, alongside the other family pets. Goodridge died just four years later. His ashes were scattered at sea near the exact spot in Penobscot Bay where he had first discovered Andre.

Andre's memory has been immortalized thanks to a larger-than-life-sized granite statue of the seal, built a year before his death. On the day of its dedication, Goodridge attached a rope to a tarp that had been draped over the sculpture, and tied the other end of the rope around Andre. He then instructed Andre to swim in the opposite direction, thus unveiling his own statue before a crowd of a thousand spectators.

The statue's placard reads, in part:

> *Andre was born on Robinson's Rock in Penobscot Bay on May 16, 1961. Abandoned at birth, he was found, befriended, raised and trained by Harry Goodridge of Rockport, Maine. Andre is honorary harbor-master of Rockport Harbor and is a celebrity of more than local renown. His antics have delighted people far and wide.*

ANDRE'S RISE TO FAME came at a time when seals were a rare sight in New England. Yet the historical record of humans and seals coexisting in the U.S. dates back thousands of years. I needed to learn more about the history of seals in the region, and what had led to their disappearance in the first place.

3

The Seal Bounty Conspiracy

WHILE SOME FISHERMEN AND others concerned about the growth of seal populations considered it to be a recent phenomenon, the seal scientists I spoke to were often quick to reference the long history of seals inhabiting U.S. Atlantic waters. The increasing numbers of gray and harbor seals, they explained, was an example of two species successfully recolonizing their former range. But if seal populations had historically thrived in New England, I wanted to understand when, how, and why they had disappeared.

Not long into my dig for answers, I came across an archaeological research study describing seal bones that were uncovered in Maine's coastal middens—piles of discarded shells that can reveal clues about ancient life. One of the researchers on the project was Donald Soctomah, the Tribal Historic Preservation Officer for the Passamaquoddy Tribe, who agreed to meet me at the Passamaquoddy Cultural Heritage Museum in Maine's easternmost Washington County.

It was a roughly four-hour drive to the museum, along quiet roads that wound through the heart of Maine's wild blueberry barrens. In the autumn, after the fruit has been harvested, the low-bush blueberry plants, which have been managed by Indigenous people since the last ice age, explode in color—brilliant shades of reds and purples like fields of fire that extend for miles.

The Passamaquoddy are one of four federally recognized tribes in Maine, along with the Maliseet, Mi'kmaq, and Penobscot. Known collectively as the Wabanaki, or "People of the Dawnland," Indigenous people in Maine have long relied on the seasonal availability of food resources. Before the arrival of European settlers, the Wabanaki hunted moose, deer, caribou, bear, and beavers in the dense forests. They harvested mussels, oysters, clams, and more in the tidal flats along the coast, and fished for salmon and herring migrating up the rivers. Ducks, loons, geese, mourning doves, and other fowl provided them with meat, eggs, and feathers, and they gathered fruits and nuts throughout the resource-rich lands.

The ocean was another important source of food for the Wabanaki. Fish were abundant, along with marine mammals—whales, porpoises, walruses, and seals. The oral histories of Native people in the region detail a long tradition of seal hunting, as seals were an important component of their diet.

But the arrival of European colonists in the seventeenth century kicked off the decline of many of the natural resources the Wabanaki relied on. From the coast up into the river valleys, settlers cut down trees to build homes and ships' masts, or to use as firewood. They destroyed entire forests to create space for farmland. They dammed rivers to power sawmills, blocking the passage of migratory fish from reaching their spawning grounds. As Native people were pushed out of their traditional hunting and fishing grounds, they became increasingly reliant on the ocean to maintain their way of life. This was particularly true for the Passamaquoddy.

The ancestral home of the Passamaquoddy, whose name comes from *pestomuhkatiyik*, meaning "people of the pollock-spearing place," extended across a vast swath of land in Eastern Maine and

Atlantic Canada. Today, the tribe has three distinct self-governing communities—two in Maine and a third in St. Andrews, New Brunswick—that comprise a tiny fraction of the area they once inhabited. And their access to coastal resources is almost nonexistent.

The Maine coastline, with its thousands of islands and peninsulas, is surprisingly vast, extending over 3,200 miles—longer even than the coast of California. But unlike California, where public beach access is built into the state's constitution, more than 90 percent of Maine's coastline is privately owned. Of that, the Passamaquoddy Tribe owns just one-quarter of a mile. That quarter-mile strip is the only coastal territory in the entire state owned by Indigenous people. People who relied on the ocean to sustain their way of life for thousands of years, who continue to rely on the ocean for its resources today, have been confined to a postage stamp of seaside property.

I ARRIVED AT THE museum just before noon and was welcomed at the door by Donald. In his late sixties with a soft-spoken, serene demeanor, Donald is a historian and archivist and serves as the tribe's Historic Preservation Officer. Among other accolades, Donald has authored several books on Passamaquoddy history, has a degree in forest management, and spent eight years as the (non-voting) tribal representative to the state legislature. He has also been involved in preserving the Passamaquoddy's natural and cultural resources, including the largest concentration of petroglyphs on the East Coast of North America.

Donald guided me through the collection of artifacts that filled the modest space—award-winning baskets woven with sweetgrass, brightly colored artwork made from dyed porcupine quills. He paused in front of a glass case featuring a bow and arrow to point out the arrow holder, which had been created from the skin of an adult seal.

Just past a stunning array of wampum belts was a display that showcased a pair of tall sealskin winter boots in near-perfect condition. Donald picked up one of the boots and gently rotated it in his hands. "They're sewn to be waterproof," he said, running his finger along a seam. "That's a waterproof stitch." The sealskin was silver with black spots. The short, fine fur was smooth to the touch, with a silken sheen. The boots looked to be about my size, with vibrant, fire-red cording stitched along the seams, and a band of longer fur from another animal sewn around the tops.

After we completed the museum tour, Donald invited me to sit in a chair beside his desk. He folded his hands on his lap, leaned back in his chair, and gazed at me, his eyes soft. "Now, what can I tell you?"

A decade or so earlier, Donald was involved in an archaeological research study led by scientists at the University of Maine who had discovered a collection of gray seal ear bones, or bullae, that had been preserved in a shell midden. The odd thing about the bullae was that they all came from the left ears of the seals, and never the right. To understand the significance of these findings, Donald began interviewing Passamaquoddy elders from seal-hunting families.

Passamaquoddy seal hunters traveled in birchbark canoes expertly built to withstand rough ocean conditions, but the journeys could be dangerous, particularly if they faced inclement weather. "They said when you harvest the seal, one bulla is left behind in the village and you take the other bulla with you," he told me. "So it means you can always come together to be one." Separating the ear bones of a single animal was a way for a seal hunter to maintain a connection to their family during their journeys. "It's sort of like a magnet," said Donald. "A spiritual magnet to come back."

One of Donald's most memorable interviews was with a tribal elder named Blanche Sockabasin. When she was young, Sockabasin's grandfather, a highly respected Passamaquoddy seal and porpoise hunter, invited her to join a seal hunt. She recounted her experience to Donald; she remembered it in vivid detail.

"They canoed around the bay to this one spot where the ledges were, where the seals like to sun themselves," Donald said. "The tide was down enough, and there was a little spit of sandy gravel where they could put the canoe in." Sockabasin's grandfather circled around to sneak up on the seal. After he killed it with his number nine buckshot, they struggled to lift the seal into the canoe due to its large size. Later, her grandfather skinned the seal, cured the hide, and processed the meat so the family could have a meal. When they finished, he carried the seal bones to the shore and released them in the water. She asked him why he put the bones back in, and he told her that he put everything back in that they didn't use. They treat the seal's bones with dignity, he explained, so the spirit of the animal can come back. "It was a spiritual connection that he had," said Donald. "He shared it with her, and then she shared it with me." Sockabasin had passed away a few years earlier. "If I didn't interview her, that story would have, you know . . ." Donald looked away, his voice trailing off.

SEALS WERE AN IMPORTANT food resource for the Passamaquoddy. But by the late-nineteenth century, that resource was under threat. Around the time of the Civil War, Maine had become a popular tourist destination, drawing summer visitors from nearby states, including Massachusetts and New York, to hotels and cottages along the picturesque coastline. As the tourism industry expanded, hotel owners and other tourist-related businesses began to see increasing value in seals as part of the pristine nature experience visitors enjoyed during their seaside vacations. But there was a problem.

Around that same time, fishermen throughout New England had grown increasingly frustrated with seals and their appetite for high-value fish. Over time, once-abundant species like cod, herring, and mackerel had been depleted. While the declines were primarily a result of overfishing, it was true that seals consumed many of the same species fishermen targeted. Since there were no laws at the time preventing the killing of seals, some fishermen took it upon themselves to eliminate their competition by shooting seals on sight. But wealthy tourists had little interest in watching seals get shot, nor did they enjoy encountering dead seals that washed up on the beaches. In the early 1870s, the tourism industry began lobbying the Maine legislature to enact seal protection laws, while the fishing industry lobbied to repeal them. The debates kicked off years of ping-ponging policies about the seals in Maine.

But by the 1880s, concerns had mounted throughout New England and across the U.S. about the status of fisheries, and fishing groups pressured state governments to take action. The governments of Maine and Massachusetts responded by placing bounties on seals—cash payments offered in exchange for a seal's nose or tail as proof of the kill. Hunters, who were most often fishermen, shot at seals from small boats in open water, earning up to five dollars for every seal killed. The hunters were asked to deliver the "proof" to the town treasurer, who was responsible for issuing the payments and then destroying the foul-smelling rotting seal parts, often by burning them.

The seal bounties were highly controversial, with many groups—not just the tourism industry but animal welfare advocates, coastal residents, and environmentalists—opposing the bounty system. Some expressed concern over the wasteful nature of the hunt, as there was no legal market for seal products in the U.S. That is, there was no commercial market. Indigenous people consumed seal meat, used sealskins for clothing and other products, and used seal oil for medicinal purposes. But the purpose of the bounty hunts was to eliminate the animals. Aside from the seal parts delivered to town clerks, the carcasses were often dumped back into the ocean.

The seal bounties took aim at one of the last remaining traditional food resources available to the Passamaquoddy Tribe. The annihilation of seals, not for human consumption but as retaliation against the simple fact that they eat fish, defied not only the foundation of Indigenous cultural beliefs, but their ecological knowledge and efforts to sustain the resources they rely on to survive. It also defied their treaty rights to hunt and fish in their traditional lands and waters.

"When they put a bounty on them, the number of seals around our waters just disappeared," said Donald. "Commercial fishermen, when they weren't fishing for cod or other types of fish, they were out sealing." Seals had been a staple resource for Passamaquoddy communities. "There was no seal at home to harvest, and you've got to still provide for your family," he said.

The tribe had to do what they could to survive. So, in response to the unprecedented slaughter of seals along the coast of New England, Passamaquoddy hunters and artisans launched one of the most creative bounty schemes in history.

IN THE FALL OF 1907, two Passamaquoddy seal hunters, John Francis and Sapiel Mitchell, arrived at the city clerk's office in Lynn, Massachusetts, with the tails of thirty seals. The men presented their collection to Lynn's city clerk, Gustavus Attwill, and pledged that they had shot and killed the seals in nearby waters. They had arrived to collect the bounty of three dollars per seal.

Attwill was dubious. The city of Lynn had been paying out less than $10 in seal bounties each year for the two or three seals killed in the Saugus River. Yet these men were claiming to have killed 10 times that number. Despite his reservations, Attwill paid the men $90 and they departed.

A few weeks later, Francis and Mitchell returned to the clerk's office, this time with 31 seal tails in tow. Attwill, now even more suspicious, questioned the two men until they eventually gave up and left without collecting a payment.

Months later, Attwill was informed that "two Indians, presumed to be the same ones" had collected $250 in seal bounties in nearby Salem, Massachusetts. Firmly convinced a scam was afoot, Attwill alerted Salem's city clerk and the county treasurer. Soon after, the Massachusetts state police launched an investigation. The prevailing theory was that the Passamaquoddy hunters were maintaining a "seal nursery" off the Maine coast—killing seals in Maine and transporting the tails to Massachusetts to reap the bounty rewards.

In early 1908, the Massachusetts state police arrived in Sipayik, a Passamaquoddy reservation in Eastern Maine, where they arrested a man by the name of Joseph Sockabasin, along with seven other tribal hunters. They charged the men with perjury and escorted them to the nearby town of Eastport, where they were held in jail. Arrangements were made to bring the men to Massachusetts by steamer ship for trial.

Days later, appearing at the Lynn city court, Sockabasin informed the judge that the tribe had secured as much as $15,000— equivalent to nearly half a million dollars today—in seal bounties in various Massachusetts towns and cities in exchange for fake seal tails. A single sealskin, he explained, when properly treated, could yield more than 150 fake tails.

The charges might have resulted in long prison sentences, but the judge was lenient. He asserted that the men deserved special consideration "because of the wrongs done their ancestors by the whites," according to local news reports. In the end, five men were found guilty, including Sockabasin, and sentenced to prison terms that ranged from two months to a year.

As word spread about the bounty scheme, people across the U.S. voiced their admiration and support for the Passamaquoddy. Some praised the hunters and craftsmen for their clever work. The *New York Sun* reported that Sockabasin ". . . has a commercial originality and a sense of humor that his white brethren do not often surpass. For the purpose of collecting a bounty on seals' tails he makes artificial seals' tails which the town clerks take gladly. This man should go into politics. He could make any platform palatable and gull any party."

There were a number of other fraudulent claims uncovered during the bounty program, submitted by non-Passamaquoddy seal hunters. Fraud has long proven to be a challenge with wildlife bounties, regardless of species or region—although fraudulent claims typically involve killing animals outside of the bounty jurisdiction as opposed to crafting artificial animal parts. But many viewed the Passamaquoddy "conspiracy" as a brilliant scheme, as well as a form of retribution.

"We lived in balance with everything," said Donald. "But once that balance was interrupted, we had to think beyond it. That's why there was a so-called conspiracy. Because it meant we didn't have to harvest as many seals."

Donald emphasized that it had also been imperative for their survival. The bounty money they received from turning in artificial seal tails enabled the tribe to feed their families without compromising their traditional values. "We grew up eating seal and porpoise meat," said Donald. "That was our only freedom to practice our traditional rights. Otherwise, we wait for the Indian agent to give us canned goods. Then we come to find out canned goods are the reason we have such a high rate of diabetes in our community."

While Passamaquoddy seal hunters had found a way to avoid killing large numbers of seals, other hunters were far more eager to reduce the populations. Between the late-nineteenth and mid-twentieth centuries, an estimated 72,000 to 135,000 gray and harbor seals were killed in New England waters as part of the bounty hunt. Even after the bounties had ended, reports of fishermen shooting seals on sight continued. By the early 1970s, gray seals had been effectively eliminated from U.S. waters, and harbor seal numbers had dwindled substantially.

After the Marine Mammal Protection Act was passed, even as seal populations began to slowly recover and despite the tribe's history of seal hunting, the Passamaquoddy, like everyone else, were federally banned from hunting seals under the terms of the law.

AFTER MY VISIT WITH Donald, I drove south to Sipayik, the second of the Passamaquoddy Tribe's reservations, to meet with another tribal member, Dwayne Tomah, at the Waponahki Museum. I arrived a few minutes early, so decided to explore a scenic lookout Dwayne had recommended.

Behind the museum, past a set of picnic tables and a collection of trees, a rock ledge extended above the Passamaquoddy Bay, offering a 360-degree view of the landscape. As I stepped toward the edge of the rock, a juvenile bald eagle, perched out of sight beneath the ledge, suddenly took flight, its broad wings thumping the air as it soared across the bay. Moments later, a small, round, glistening head popped out of the water a short distance from where I stood—a harbor seal. The seal spun in place as it looked around, at which point another head appeared. The two seals balanced their heads above the water as they swam toward the mouth of the bay, leaving a trail of ripples in their wake.

In his late fifties, Dwayne is the youngest fluent speaker of Passamaquoddy and the tribe's official language keeper. But his "colonial titles," he told me with a smile, were director and curator of the Waponahki Museum.

As we sat across from each other at one of the picnic tables near the lookout, I was captivated by Dwayne's rhythmic way of speaking. He seemed to enunciate each syllable with intention, and often repeated words and phrases to emphasize their importance while bouncing seamlessly between English and Passamaquoddy. "*Ahkiq*," he said. "I'm sorry?" "*Ahkiq*," he repeated. "It means 'seal' in our language."

Midway through our discussion, Dwayne offered to give me a driving tour of Sipayik. We hopped into his truck and drove across the reservation, which didn't take long, as the plot of land was less than a square mile. It was a stark reminder of the atrocities Wabanaki people have faced and the ways their access to their traditional lands and resources has been repeatedly cut off.

In the 1970s, the Passamaquoddy Tribe asserted its claim to tribal land that had been illegally sold by the state of Maine—land that amounted to almost two thirds of the state. Federal courts affirmed the tribe's claim, since the state's land transfers and sales had never been approved by Congress. But years of legal battles and intense political pressure eventually resulted in the 1980 federal Maine Indian Claims Settlement Act, which provided a reparations payment in exchange for the land, but also gave Maine a level of jurisdiction over tribes not found in any other state.

According to a 2022 study published by the Harvard Kennedy School, between 1989 and 2020, the economies of the Wabanaki

Nations in Maine grew at a rate of just 9 percent, a fraction of the 61 percent growth experienced by all other federally recognized Native American tribes, and less than half of the overall U.S. growth rate of 25 percent. The study's lead author, Joseph Kalt, and his colleagues attributed this stunted growth to the Maine Indian Claims Settlement Act, which has prevented the Wabanaki from accessing some of the federal resources available to all other tribes in the country. Yet the state has repeatedly shut down the tribes' efforts to reclaim their sovereignty.

Despite the continued battles the Wabanaki face to secure their rights in the state, Dwayne seemed to possess a wealth of energy and optimism. "We're still here," he said with a smile. "We're trying to *educate* our oppressor. Isn't that amazing? To educate your oppressor?" "It must be exhausting," I said. It wasn't lost on me that both he and Donald had taken the time to patiently share their knowledge with me that afternoon. Dwayne nodded. "Very exhausting," he said. "But we're doing it, and what does that tell you about our resiliency? What does it tell you about our *spirit*? The spirit of our people?"

As we drove past the waterfront, along tidal flats where Passamaquoddy clammers have dug for bivalves for millennia, I asked Dwayne about the tribe's relationship with seals. He seemed confused by the question. "Our people were fishermen," he said. "And the *relationship* we have with all of the animals, not just with the seals, is consistent." The Passamaquoddy wanted to provide for themselves and for their families, said Dwayne, but they never wanted to deplete the resources. "They understood that it's *give*, and *take*," he said. "It's give, and take, to be able to *sustain*, and *manage*."

AS I DROVE HOME that evening, I thought about one of the stories Donald had shared with me from his interview with Blanche

Sockabasin. When she was young, her grandfather came across a seal pup whose mother appeared to have abandoned it. He wasn't hunting seals at the time, but he suspected the pup's mother had died and was reluctant to leave it on its own. He knew that would mean a certain death for the young seal. So he lifted the pup into his canoe and brought it home, to his granddaughter's delight. Sockabasin and her family raised the seal until it was big enough to release.

"She said that seal pup was like a dog," Donald told me. "When she'd walk to school, it would try to follow her, and she'd chase it back." When her grandfather told her it was time to let the seal go, Sockabasin told Donald that she had cried and cried. But she watched her grandfather release the seal into the ocean. "I just love that story," said Donald, a rare smile lighting his face.

4

The Mimic

DESPITE THE SURPRISING NUMBER of stories I'd amassed about close interactions between humans and wild seals, for the vast majority of us, our experiences with seals tend to occur in captive settings, in zoos and aquariums.

Nearly forty years after the tenures of Hoover and Andre, the talking and traveling seals, the harbor seal exhibit at the New England Aquarium, overlooking Boston Harbor, remains in the same location in the aquarium's front plaza, where visitors and passersby can watch the seals as they swim, dive, pirouette, porpoise, and bob. The 42,000-gallon, open-air saltwater tank is surrounded on three sides with plexiglass walls, with the fourth wall connecting the enclosure to the aquarium's main building. Replicas of rocky outcrops are scattered throughout the exhibit, reminiscent of the seals' natural habitat along the coast of Maine, including flat ledges where the seals can haul out of the water to rest and interact with their human trainers.

In the final days of autumn, I drove south from Maine to Massachusetts, excited to meet the aquarium's fin-footed residents. As I approached the entrance to the building, I paused to watch the seals in their exhibit. Their underwater agility was captivating, a far cry from their clumsy caterpillar shuffles on land, and I crouched down beside the glass wall for a better view. Several seals jetted through the water, their sleek, streamlined bodies bending and twisting as they rounded the corners of the tank at high speed, then dove, one after the other, beneath a faux rock formation. A lone fifth seal had opted to forgo the morning aerobics. *A seal after my own heart*, I thought, as I watched it bob lazily in the corner of the tank. Its head, nestled in the blubbery folds of its neck, slowly rotated until the seal was facing me. Fixing its large eyes on mine, it began to drift forward. I was transfixed by this wide-eyed, curious creature, which I assumed would stop short before nearing the glass wall where I crouched. Instead, in a comically slow overshoot, the seal drifted snout-first into the wall, its nostrils and whiskers gently smooshing against the glass in a perfect blubber boop. It then closed its eyes, as if mildly embarrassed, retreated a few inches, and reopened them as it continued to stare at me. A young girl walked up behind me, so I stood and ushered her forward while continuing on to the visitor's entrance.

Once inside, I was greeted by Sean Rothwell, a senior pinniped trainer who cares for the aquarium's Atlantic harbor seals, California sea lions, and northern fur seals. In his early thirties, with reddish-brown hair and a trim beard, Sean was wearing a light jacket and tan shorts, despite the 40-degree temperatures that morning. He led me past the visitor's entrance, through the room that houses the touch tank exhibit—temporarily roped off for renovations—to the back entrance of the seal exhibit. The door opened into a small prep area where several trainers were already gathered. They were filling metal buckets with fish, the type and cuts prepared according to the preferences of the individual seals. Sean handed me a pair of tall black rubber boots to swap for my sneakers, and then began preparing his own bucket—a combination of herring, capelin, and squid.

As soon as the fish was ready, I followed Sean through a small entryway that led into the seal enclosure. A mild fish scent wafted through the cool, salty air of the exhibit. The seals sped through the water like torpedoes, occasionally popping their heads above the surface as they watched us filter in. Sean directed me to kneel on a rubber mat at the edge of the tank, while the other trainers spread out across the exhibit to assume their designated positions, metal buckets in hand. A few spectators had gathered along the outer edge of the glass walls where I'd stood just minutes earlier.

TODAY, MOST PINNIPEDS IN captivity, including harbor seals, come from captive breeding programs or from stranding and rehabilitation programs, in cases where they're unable to return to the wild. In 2021, for example, Marine Mammals of Maine responded to a harbor seal pup off the coast of Kittery that had been blinded after being struck in the head by a boat propeller. The seal, which the team named Kitt, was sent to the Woods Hole Science Aquarium in Massachusetts, where she quickly bonded with the aquarium's only other resident harbor seal, the twenty-two-year-old, captive-born Bubba, occasionally hitching rides on Bubba's back. (Sadly, Kitt died about a year after her arrival for unknown reasons, although possibly resulting from seizures caused by her prior head injury.)

The first captive seals date back to the Roman Empire. The Romans' obsession with exotic wildlife led to hundreds of thousands of animals—from elephants and crocodiles to bears, boars, and bison—being rounded up across the globe for the purpose of entertainment. The animals were trapped, snared, and forced to endure arduous journeys along supply routes to face their ultimate demise before bloodthirsty crowds in the arena.

Mediterranean monk seals and other pinnipeds were among the wildlife targeted. In one account, detailed in a 1999 paper authored by William Johnson and David Lavigne of the International

Marine Mammal Association, Roman emperor Nero ordered the flooding of an arena so he could watch polar bears catching seals. "It is likely that the monk seal entertained the crowds at many of these gory tributes to the might of Rome, yet one can only guess at the number of individual animals that met their deaths in the sand-filled arenas," they wrote.

Monk seals were also trained to perform for the public in traveling circuses and parades. The seals were highly intelligent and could easily be trained to perform tricks, including "to salute the public with their voice . . . and when called by name to reply with a harsh roar," according to a passage written by Pliny the Elder in *Natural History*. The seals continued to appear in traveling circuses in Europe into the 1800s, advertised as mermen, sea serpents, and "talking fish."

Over the past century, the number of zoos and aquariums displaying pinnipeds and other marine mammals has increased to meet growing demand from the public, particularly in North America, Europe, Australia, and across parts of Asia.

THE FIVE HARBOR SEALS residing at the New England Aquarium on the day of my visit had all been born at the aquarium. The seals ranged in age from the twenty-seven-year-old Chacoda to his thirty-seven-year-old mother, Trumpet. But even Chacoda, the youngster of the group, was considered geriatric by seal standards. In the wild, the life-span of a harbor seal is thought to be about twenty-five years, but in captivity, in the absence of predators and with the help of high-quality veterinary medicine, seals can live much longer. Barney, a harbor seal "super senior" at the Seattle Aquarium, celebrated his thirty-eighth birthday in 2023, complete with an ice sculpture of a cake and 38 frozen fish "candles." But the oldest living harbor seal on record, Skinny, is forty-eight and counting—older than many of her human caretakers at the Oregon Coast Aquarium.

Sean sat beside me on the mat, placing his boots in the shallow water on the flat ledge in front of us, as we watched the seals jet past. Sean is originally from Wichita, Kansas, thousands of miles from the coastal habitats of the flippered creatures he now spends his days caring for, but he said he's always wanted to work with animals. After college, he applied for an internship with the New England Aquarium. From then on, he was hooked.

"You can see how they move their hind flippers side to side to propel themselves," he said. "They really only use their front flippers to change direction, and also for balance when they slug around on land."

One by one, the seals hauled their smooth, plump bodies out of the water beside their trainers. I watched with fascination as the seals on the opposite end of the exhibit assumed their feeding positions, when suddenly, directly in front of us, Chacoda emerged from the depths, expertly sliding onto the ledge before us like a baseball legend crossing home plate. My heart was racing as I realized I was kneeling just a few feet from a 170-pound harbor seal.

Chacoda, for his part, was calmly assessing the situation. He looked at Sean, then looked at the metal bucket positioned between us, then looked at me, the newcomer, then back at the bucket, his snout softly twitching. I couldn't take my eyes off of him—it was the closest I'd ever been to a seal. I'd assumed my unfamiliar presence might alarm him, but he seemed far more focused on which of us humans would be the first to offer him a snack. Seeing Chacoda up close, I was surprised by the size and thickness of his whiskers—less like the fine whiskers of a cat and more like thick, sturdy porcupine quills.

Scientists have only recently begun to understand the role of seals' whiskers when it comes to navigation and hunting. In the early 2000s, to assess the power of this "sixth sense," German scientists at the University of Rostock went so far as to place a blindfold and earmuffs on a harbor seal to study the role of its highly sensitive vibrissae. They determined that the animal, in the absence of its other senses, could detect the presence of moving objects using only its whiskers.

At the New England Aquarium, Chacoda and his pool-mates were involved in whisker research of their own. Several years back, a team of scientists from MIT's Center of Ocean Engineering visited the harbor seals at the New England Aquarium to learn more about the design and mechanics of seal whiskers. When I first heard about the research, I wondered why a group of marine engineers was interested in the biology of harbor seals. But I soon discovered that the study was part of a broader scientific effort known as "marine biomimetics"—the process of using features observed in nature to inform the design of ocean technologies.

The MIT researchers, led by Heather Beem and Michael Triantafyllou, wanted to understand exactly how seal whiskers functioned. To investigate, they 3-D printed giant whisker replicas—fifty times larger than real ones—which they dragged through the water while exposing the whiskers to vibrations at different frequencies. When an object like a fish travels through water, it creates a trail of tiny whirlpools with each flick of its fin, leaving a set of temporary underwater footprints in its wake. When this external disturbance occurs, even from a distance of forty meters, seals' whiskers, with their unique, wavy shape and wealth of nerve cells, begin to vibrate at exactly the same frequency. The vibrations help the seal to determine the size, shape, speed, and even direction of the moving object. Seals' whiskers can also differentiate between their own movements and those of an external object, enabling them to swim through the water while stealthily zeroing in on their prey.

The team is now studying ways to replicate the design and mechanics of seals' whiskers to build technologies for ships and underwater vehicles to better "sense" their environment, as well as whisker-like sensors for marine robots that could detect fish, marine mammals, ships, and submarines, or even to help to identify sources of pollution, such as the spread of an oil spill. By mimicking seals' unique biological functions, we're expanding our own capabilities as humans.

Sean threw Chacoda, or "Chucky," as the staff affectionately called him, a small piece of herring, which he deftly caught in his thick-whiskered snout. "Did you know harbor seals have a split tongue?" Sean asked. Before I could respond, he looked at Chucky and said, "Tongue!" in a sharp, commanding voice. Chucky immediately opened his mouth wide and stuck out his tongue, patiently holding it in place as Sean pointed out various areas of interest inside his mouth. The tip of Chucky's tongue was indeed split in two like the prongs of a fork, which is believed to help seals trap and manipulate their prey, and his mouth was filled with sharp, pointy teeth. Seals typically consume prey whole, using their teeth to hunt and grab, or for slicing fish or other prey that's too large to consume in a single gulp. They have no need for flat molars like humans. Chucky's teeth were in great condition thanks to his daily tooth brushings. The chompers of aging seals are prone to the same tooth decay as humans, so the trainers are careful to spend time caring for the seals' teeth. Each animal has a personalized bamboo toothbrush with his or her name on it.

"Good job, buddy," said Sean. Chucky retracted his tongue as Sean threw him another fish. "All right, now I'll ask him to do a vocalization." He turned to Chucky, placing his hand gently under the seal's chin, and said, slowly, "Listen." Chucky stared intently at Sean. "Hey!" said Sean. In response, Chucky thrust his neck forward like a peacock while repeating the word in a surprisingly deep, gruff voice: "Hey!" "Hey!" said Sean again. "Hey!" Chucky repeated, his neck again lurching forward. "Good," said Sean, throwing him a piece of fish. Sean placed his hand back under the seal's chin. "Listen," he said. Chucky watched Sean, motionless. "How. Are. You," said Sean, enunciating each word in clear, staccato commands. Chucky drew his head back, opened his mouth wide, then slowly stretched his neck forward as he repeated the phrase, "Hooow-rrrr-ooooo," the three words melting into what can best be described as a vowel howl.

By this time, I'd begun to feel a mild pain in my cheeks, and I realized I hadn't stopped smiling since the moment I'd stepped foot in the seal enclosure. It defied logic that I was sitting mere feet from a harbor seal capable of speaking to me in plain English. While Chucky's vocal talents were limited to just those two expressions, along with a series of sounds Sean referred to as "random vocals"—a ten-second burst of burps, grunts, and screeches Chucky emits in a single breath—it still seemed remarkable. He sounded so . . . human.

Chucky is the first seal at the New England Aquarium since Hoover able to mimic human speech, which maybe isn't so surprising, since Chucky is Hoover's grand-pup. While none of Hoover's immediate offspring exhibited any ability to mimic human speech, his grand-pup has carried on his legacy.

A FEW MONTHS BEFORE my visit, a team of Dutch researchers from the Max Planck Institute for Psycholinguistics, equipped with high-tech audio devices, visited the New England Aquarium to observe Chacoda and record his vocalizations. The team, led by Diandra Duengen, aimed to compare the recordings against

Hoover's. For years, the scientists had been studying the recordings of Hoover's vocalizations as part of their research on vocal production learning.

Extremely rare among mammals, vocal production learning is the ability to create or alter sounds based on interactions with others—the way humans eventually learn how to speak by hearing and repeating the sounds and words of a language, or languages, spoken to them from birth. Most of the species known to possess this ability are birds, along with a small number of mammals, including humans, seals, whales, dolphins, bats, and elephants.

According to Duengen, Hoover is one of the best-documented examples of vocal-production learning when it comes to human speech that has ever been exhibited by a mammal. His recordings have helped linguists to understand the close evolutionary ties between humans and pinnipeds. As Duengen wrote in a 2023 article published in the scientific journal *Current Biology*, "Hoover may eventually be immortalized as one of the godfathers of mammalian vocal production learning research."

IN ADDITION TO BEING a senior pinniped trainer at the aquarium, Sean also serves as the pinniped enrichment coordinator, which, as you might guess, means he is responsible for creating games and activities to engage not just the harbor seals, but also the sea lions and fur seals, which live in a separate exhibit toward the rear of the aquarium.

Pinnipeds are highly intelligent, curious creatures, and the activities and devices Sean creates, which range from floating toy buckets with various levers and hoops to object-retrieval games, are designed to improve animal welfare by encouraging the animals to solve problems, forage, and play. The activities keep the animals engaged while stimulating their natural behaviors. This type of cognitive training and enrichment helps to protect against tedium and stress. The trainers also keep a close eye out for any signs of "pattern swimming," where captive animals engage in a cyclical pattern of movement, believed to be a sign of boredom.

One of the most famous cases of this was a polar bear named Gus at the Central Park Zoo in New York that would swim in repetitive figure-eight patterns for up to twelve hours per day. Eventually, an animal behaviorist (basically a polar bear therapist) concluded Gus was bored, so zookeepers developed an extensive enrichment program. They redesigned Gus's habitat and created a variety of toys and challenges to entertain him. Within months, Gus's pattern swimming declined considerably, although it never fully disappeared.

After Chucky had completed his vocalizations, Sean gave him a new command: "Bubbles!" Chucky placed his snout in the shallow water and blew air out of his nostrils, looking up eagerly at Sean when he was done. Sean threw him a fish, then said, "Porpoise!" Chucky immediately slid off the ledge and disappeared under the water. I held my breath, with no idea of what came next. Seconds later, Chucky leapt out of the water in a stunning display of streamlined agility, an act he repeated twice more before returning to the ledge for his fish reward.

"Do you want to touch him?" asked Sean. I nodded enthusiastically and Sean guided me to gently stroke the back of Chucky's head. His skin was slick but not slimy in the way I'd expected. I could feel the very short coat of fur that covered his thick blubber. "Do you want a kiss?" said Sean. I looked at Chucky, waiting for him to react to what I assumed was a training command. When nothing happened, I quickly realized Sean was staring at me. "Wait, me?" I asked. "Are you serious?" Sean laughed and talked me through the commands.

First, I presented my hand in a fist and said, "Mark." Chucky slid toward me and lightly placed his mouth on my hand. "Good," said Sean, giving him a small piece of fish. "Now," he said, "Put your face forward for a kiss. Just point to your face and say, 'Mark.'" My heart racing, I leaned my face out over the ledge, pointed to my mouth and said, "Mark." Chucky swam up and planted his cold, wet, fishy mouth right on mine, patiently holding his pose. Holding my own pose, I wondered, how do you know when a seal kiss is over? Should I pull away? Would Chucky pull away? Since this might be my first and last seal kiss, there was no way I would be the first to end it, so I held my mouth in place for what felt like a second or two beyond what might be socially acceptable, until Sean softly cleared his throat and said, "Okay, that should do it." Sean offered to let me do it once more, gently adding that people mostly point to their cheek, rather than their mouth, in case I wanted to go that route. I regret nothing.

I'D HOPED TO TAKE advantage of my visit to the aquarium to also learn more about Hoover. While I had tracked down a couple of recordings of his vocalizations from the 1980s, the sound bites were each only a few seconds long and the audio quality was limited, so it was hard to get a sense of exactly how "human-y" his voice sounded. Thankfully, Patty Schilling, the aquarium's marine mammal supervisor, offered to let me listen to a compilation of Hoover's audio clips the aquarium had on file.

I met Patty in her office in the aquarium's main building, not far from the 200,000-gallon, cylindrical Giant Ocean Tank (where I paused for a quick glimpse of one of the aquarium's most famous residents—a nearly one-hundred-year-old, 550-pound green sea turtle named Myrtle). Patty first began working at the aquarium in 2003, nearly twenty years after Hoover's death, but she had heard stories about the charismatic talking seal from prior trainers. She invited me to sit in a chair beside her desk as she clicked through her files to find the recordings on her computer. "Ah, here they are," she said, queuing them up.

A deep, raspy voice suddenly filled the room, firing off words in rapid succession. "Hey, hey, hi, hey, hey, hey, he-ey." I couldn't believe it. "Ya-ya-ya-ya-ya-ya-ya! Hey, ya! Hey-oo!" My mouth hung open; I was flabbergasted. Had I not known these were the sounds of a seal, I would have bet a considerable sum of money that the voice was coming from a half-drunk, older gentleman with a penchant for smoking Marlboro Reds. "Hoo-vah, Hoova, Hoova, hoo-VAH, hoo-VAH, Hooo-vaaa." Patty and I leaned back in our chairs simultaneously as she continued playing the recordings, as though we were listening to a jazz album, instead of the gruff, bass vocalizations of a harbor seal. "Get ovah they-ah, come hee-ya, Hoo-VAH, get ovah hee-ya, get outta they-ah, Hoo-vah, Hoo-VAH!"

We were both laughing by the time Patty reached the end of the recordings. "You can imagine the dynamic between this animal and that person," she said. "How interesting of a relationship they must have had."

While Patty had never met Hoover, she had spent the last twenty years working with his grand-pup. When Chucky first began to vocalize as a young seal, his trainers worked closely with him to encourage his speech, rewarding the odd noises he made with fish. By the early 2000s, Chucky was frequently vocalizing all on his own. "He would just sit in the corner of the exhibit and holler," she said. "At night, you'd be walking down the plaza and you could hear him vocalizing. It sounded like a demon coming from the building." As he got older, he vocalized on his own less and less, likely as a result of decreased testosterone levels, but he continued to speak on command.

While most true seal species tend to stay close to their pups while nursing, harbor seals are an exception, and the pups' unique vocalizations are key to their survival. Unlike most seal species, which are born with a thick fur layer called "lanugo" that they must shed before entering the water, harbor seal pups shed their lanugo in utero, and are therefore able to swim within hours of birth. Even so, they're dependent on their mothers' milk until they gain the weight and strength needed to begin foraging on their own. During this nursing period, harbor seal mothers will leave their pups on rocks or beaches for extended periods while they go off to hunt. When they return, they rely on their pups' distinctive calls, which can be heard for more than half a mile, to find them.

I asked Patty what the biggest challenge was that she and her team of trainers faced in working with harbor seals.

"The language barrier," she said immediately. Her answer surprised me, given that we had just been listening to the recordings of a talking seal. But there's a difference between mimicking human speech and understanding human language—the ability to extract meaning from words and phrases. If a seal wasn't feeling well, or if it didn't respond to certain training efforts, it was up to the trainer to identify the root cause. Despite Hoover and Chucky's talking talents, their thoughts and emotions were unknowable to us.

For hundreds of years, language has been defined as one of the key differences that separates us from the rest of the animal kingdom. Yet now, animal communication experts are discovering more about the language learning abilities of other species—and the ways in which our evolutionary pasts may be intertwined. While Hoover and Chucky's gift of mimicry doesn't quite reach Shakespearean levels of sophistication—perhaps closer to a guy selling Fenway franks at a Red Sox game—understanding the mechanics behind their ability to hear, process, and repeat specific sounds is helping us to better understand the evolution of language learning in humans as well.

When Hoover first began vocalizing, he was the first mammal known to science at the time to mimic human speech. Since then, scientists have observed and recorded other mammals with similar capabilities, including whales, dolphins, gorillas, and elephants. And we've learned even more about the remarkable vocal learning capacity of seals. In 2019, researchers at the University of St. Andrews taught a gray seal named Zola to "sing" the first few notes of the Star Wars theme music, along with "Twinkle, Twinkle, Little Star" (although "singing" was a generous term, as she sounds a bit like a honking goose).

Bruce Moore, a professor in the Psychology Department at Dalhousie University in Halifax, Nova Scotia, was fascinated by Hoover's vocal abilities. In the book *Social Learning in Animals*, edited by Cecilia Hayes and Bennett Galef, Moore authored a chapter on imitative learning and wrote that he once played a tape of Hoover's vocalizations for his students. "When the seal laughed, we laughed so loudly that the parrot in the next room heard us," he wrote. "It then joined in, and therefore, for perhaps the first time ever, three very different species laughed together."

Moore wondered whether seals had a similar capacity for mimicking not just human vocalizations, but human movements as well. To test this theory, he temporarily adopted a neonatal harbor seal, which he named Chimo after an Inuktitut greeting. Chimo quickly imprinted on Moore and his two Dalhousie colleagues, following them everywhere and nuzzling their feet in an attempt to nurse. While the seal was never observed mimicking human behavior, he did demonstrate sexual imprinting after making several "unambiguous" advances to a young woman. (Moore was careful to note that the seal's approaches were both "unprovoked and unrequited.")

NOT LONG AFTER MY journey to the New England Aquarium, a new acquaintance, Bill Barton, invited me to visit his family's vacation home in Maine. Bill was a skilled sailor from Massachusetts whom I'd met through a mutual friend. When he learned about my seal project, he suggested I meet him and his wife, Annie, during low tide to observe the harbor seals that haul out on a nearby ledge.

By the time I arrived, a dozen seals had already gathered, positioned side by side on the shallow rock, with several exhibiting their trademark banana poses. As we watched the seals, I mentioned my recent visit to the New England Aquarium to meet the resident harbor seals. Bill's eyes suddenly lit up. He asked if I'd like to hear his "craziest seal story."

"Absolutely."

In 1977, Bill joined a research team studying the behavior of humpback whales aboard a three-masted sailing ship, *Regina Maris*. The ship sailed from the whales' breeding habitat off the Dominican Republic to their feeding grounds off the northern coast of Labrador. As they moved north, the ship docked at various ports along the eastern seaboard, and crew members would take turns monitoring the deck throughout the night to prevent any curious pedestrians from inviting themselves aboard.

A few years later, the *Regina Maris* was docked in Boston Harbor and Bill was on watch duty. With the crew fast asleep belowdecks, and not a single pedestrian in sight, he decided to keep himself awake by taking a short walk along the wharf. The moon was bright that evening, the water still, and Bill gazed up at the historic brick buildings that overlooked the harbor, breathing in the quiet peace of a sleeping city. Suddenly, a loud, gruff voice shattered the silence. "Hey you," the voice said. "Get outta there!" Bill wheeled around to face his aggressor. He assumed the man must be drunk given his slightly slurred speech, but there was no one there. Confused, he spun in a circle, his eyes darting between dark shadows as he tried to figure out where the man was hiding.

As he turned back toward the city, the man called out again, his voice deep and raspy, "Hey you. Get outta there!" Bill spun around again, but there was still no one there. He was baffled. As far as he could tell, the only creatures still awake along the harbor walk that evening, besides him, were a few seals in an outdoor tank in front of the New England Aquarium. He watched the animals in their enclosure as he tried to get his bearings, but then something unexpected happened. One of the seals, bobbing vertically in the water while facing Bill, slowly leaned his head back and called out, in clear English, "Hey you, get outta there!"

By the time Bill returned to the ship, he'd decided to keep the story to himself—after all, who would believe that he'd met a talking seal? He was sure the crew would chalk it up to late-night lunacy. But the next morning, when he still couldn't make sense of the encounter, he returned to the aquarium and informed a staff member about what had happened, bracing himself for the inevitable look of disbelief. Instead, the staffer smiled knowingly and said, "It sounds like you met Hoover."

BEFORE LEAVING THE MUSEUM, I stopped at the aquarium's gift shop to see if I could pick up a few souvenirs for my toddler nephews. While there weren't too many seal-themed gift options, I did come across some fascinating and very original artwork. In recent years, Patty, Sean, and their team had introduced a new type of seal enrichment activity. The trainers had taught several of the seals, including Chucky, to hold a paintbrush in their mouths, dip the brush on a paint palette, and move it back and forth across a blank canvas held by one of the staff. The seals' preferred medium was, of course, watercolors.

There was a chill in the air as I walked back to my car, and I wrapped my scarf tightly around my neck. Just outside of the museum, I paused by an overlook that offered visitors a picturesque view of the Boston Harbor. How many harbor seals were out there right now, I wondered, swimming and twirling beneath the surface of the murky water?

Gray seals, meanwhile, had plenty to keep them busy. In just a few weeks, pregnant females would begin hauling out in large breeding colonies in Massachusetts, Maine, and Atlantic Canada to give birth to and nurse their pups. The males would come ashore soon after to mate with the females. To withstand the weeks of fasting required during the breeding season, gray seals needed to consume as much food as they could before they hauled out. I had heard and read about the chaotic spectacle—massive numbers of seals gathered in close proximity, howling alongside their white-coated pups. With winter just around the corner, I was determined to find a way to experience the scene for myself.

PART TWO
WINTER

There is a little bit of every season in each season...The calendar, the weather, and the behavior of wild creatures have the slimmest of connections. Everything overlaps smoothly for only a few weeks each season, and then it all tangles up again.

—ANNIE DILLARD, *PILGRIM AT TINKER CREEK*

5

Why Did the Seal Cross the Road?

ON A FRIGID JANUARY morning, I woke hours before the sun, bid my sleepy dogs farewell, and headed out into the blue-black winter darkness. It was an hour's drive east to Rockland along quiet coastal roads. I arrived just before six a.m. and fell in line behind a small parade of men silently marching toward Journey's End Marina.

The tide was nearly out as we approached the *M/V Equinox*, a forty-foot winterized transport vessel. Originally a sea ambulance for Monhegan Island, the *Equinox*—which resembled a lobster boat with the added feature of an extended, heated cabin—was now used to shuttle folks from the mainland to various offshore islands in Maine. Over the years, the boat's captain, John Morin, has transported an eclectic mix of passenger groups, from U.S. Navy officials to scientists, federal wildlife managers, and tour operators. That morning, he was ferrying two dozen construction workers and one enthusiastic seal observer—me.

As the *Equinox* motored past Rockland's Breakwater Lighthouse, a subtle orange glow began to take shape along the horizon. Heading east into open water, Captain John steered us toward our first stop, North Haven Island, where the construction crews would begin their shifts. Sea spray smashed against the windows of the cabin, leaving behind a kaleidoscope of ice crystals that washed away and re-formed with each breaking wave. Thick icicles held fast to the bowlines like rows of translucent sharks' teeth.

Minutes before we reached North Haven, sunlight broke through the darkness, illuminating a thick sheet of sea ice across the bow. One by one, the workers disembarked, nodding their thanks to John on the dock. Around seven thirty a.m., we continued south, meandering through the calm, protected narrows along the coast of Vinalhaven toward our final destination: Seal Island.

Located over twenty miles offshore, Seal Island is the most remote island in Maine. A haven for nesting seabirds, the island is a national wildlife refuge managed by the U.S. Fish and Wildlife Service. In recent months, I'd discovered that Seal Island was home to the largest breeding colony of gray seals in Maine. While I'd frequently seen harbor seals hauled out along the coast of Maine, I had yet to observe any gray seals. Winter pupping season would be my best opportunity to see them on land. The challenge was, during pupping season, gray seals tended to gravitate to remote, hard-to-access islands and beaches. Seal Island was an ideal location for seals, partly because humans were forbidden from stepping foot on its shores without a federal permit . . . and for good reason.

In the late 1970s, a mysterious fire broke out on the island, and a group of makeshift firefighters was called in to fight back the flames—a mix of students and staff from Outward Bound, a wilderness school located on nearby Hurricane Island, as well as local foresters and state prisoners. They used salt water and brushfire pump tanks to control the blaze, but the peat beneath the grass continued to burn.

Days into the effort, there was a sudden explosion, and shrapnel began flying off the island. An Outward Bound staffer quickly radioed the course director to describe what had happened. "What kind of explosion is it?" the director asked. "The kind that tears limbs from torsos," he replied. The crews were quickly evacuated.

Seal Island is shaped like a handlebar mustache, a squiggly line in the sea, or, for the purposes of U.S. Navy bomber pilots during World War II, an enemy ship. For years, the Navy had used the island as a bombing target for training exercises, and unexploded ordnance was buried beneath its surface. Today, due to ongoing safety concerns, the island remains uninhabited, and visitors are banned from stepping foot on its shores without federal authorization. Between its remote location and explosive personality, Seal Island was an odd destination for a fifteen-degree January morning. But its inaccessibility to humans is a big part of what makes the island particularly well-suited to a breeding colony of gray seals.

We approached Seal Island from the northwest. When we were about two hundred feet from its rocky shoreline, Captain John cut the motor. The island rose like a staircase from the sea, overlapping layers of granite slabs speckled with shiny gray and brown rocks. Overhead, hundreds of gulls circled and squawked.

"Look at all of those seals," said John. I turned to look at him to see if he was joking. As far as I could tell, there wasn't a single visible animal on the island other than birds. But as I stared at the ledges, some of the large, shiny rocks began to move. Suddenly, scores of seals, their dark, mottled coats the same grays and browns as the island itself, began to take shape, as though I'd just solved a Magic Eye painting. As we stood on the deck watching them, a cacophony of eerie moans, like the songs of a haunted choir, drifted toward us from the island, carried by the wind.

I was amazed not only by the sheer number of seals on the island, but by the enormous size of the animals. By that time, I'd grown somewhat accustomed to harbor seals, which typically range in size between two hundred and three hundred pounds, but gray seals were another story. While the females weigh about four hundred pounds on average, the males can weigh upward of nine hundred pounds—about the size of a grand piano. The females were gathered mostly along the highest ledges on the island, with fuzzy, white-coated pups huddled closely beside them. The males, meanwhile, were scattered across the lower ledges, closer to the surf, biding their time until the females finished their intensive nursing period and were ready to mate.

The pupping behavior for gray seals is quite unlike that of harbor seals. Harbor seal pupping occurs in the late spring and early summer months, at which point pregnant females individually seek out a quiet spot along the coast to give birth, then nurse their pups (who can swim just hours after they're born) for four to six weeks while teaching them to hunt and forage. Gray seal pupping, on the other hand, occurs in the brutal dead of winter. Between late December and mid-February each year, female gray seals gather in large colonies to give birth, nurse their pups for about three weeks, and then promptly mate again before returning to their mostly solitary lives in the water. The pups are left alone on the island to survive off their fat reserves for a few weeks before they, too, take to the water.

As the *Equinox* gently bobbed in the waves, two males positioned nose-to-nose on a ledge suddenly rose up, growling and hissing as they faced off, smashing their hulking frames into one another. The smaller of the two was knocked from the ledge, and fell several feet onto a granite slab below. Shockingly, the seal appeared unfazed by the fall, likely cushioned by his thick blubber layer, and immediately lifted his head toward his nemesis above as he continued growling.

Meanwhile, several of the other males, which had been eyeing us reproachfully as we drifted along the western shore, plopped into the water. Seconds later, we heard loud, wet releases of air, like whales exhaling through their blowholes, as multiple sets of enormous nostrils emerged from the depths just feet from the boat. Unlike the cute, doglike snouts of harbor seals, complete with their heart-shaped nostrils, male gray seals have donkey-like snouts, with nostrils shaped more like a giant moth. Their distinct schnozzles have earned them the nickname "horseheads." The scientific name for gray seals, *Halichoerus grypus*, is even less flattering. It translates from Latin to "hook-nosed sea pig." (The scientific name for harbor seals, on the other hand, is far less offensive—*Phoca vitulina*, or "calflike seal.")

Several bald eagles circled above the southern tip of the island, where vertical rock cliffs at least fifty feet high dropped like a ship's hull into the water. One of the eagles was clutching what looked to be a long, red, stringy mass in its talons. "Seal placenta," John explained. Decades earlier, a few eagles started showing up on the island during seal pupping season, the nutrient-rich delicacy luring them twenty-two miles offshore. Eventually, word spread about the bounty of seal placenta on the island, and the raptors started appearing in droves every January, timing their arrival with that of the newborn pups.

Bald eagles had made their own remarkable comeback in recent decades. I had to imagine this population of eagles that wintered along the Maine coast was quite happy with the increasing seal numbers. But I wondered what else was known about the ways in which seals impact other species in their environment. While the role of seals as fish-loving predators, not to mention the beloved prey of great white sharks, had received considerable attention over the years, there had been much less focus on their broader functions within their marine habitat. Yet limited available data suggest that the ecological impacts of seals, as well as other pinnipeds, extend far beyond what they eat.

In a 2016 analysis published in the *Annual Review of Environment and Resources*, James Estes, a pioneering marine ecologist, and his coauthors identified just five published studies that focused on the impacts of pinnipeds on ocean ecosystems. Despite the small sample size, the findings suggest that pinnipeds can influence their environment in significant ways. In Antarctica, for example, the mere presence of leopard seals can impact where and how penguins forage. I was particularly interested in the concept of trophic cascades—the indirect effects of predators at higher trophic levels that can trickle down through a food web.

In trying to understand the phenomenon, I came across an illustrative story about sea otters. In the early 1970s, Estes was studying the impacts of sea otters on kelp forests in the North Pacific. He discovered that in areas where otters were abundant, kelp forests were thriving, while in areas where there were few otters, he found very little kelp and, surprisingly, a lot of sea urchins. As it turned out, sea otters were indirectly supporting kelp forests by consuming large numbers of sea urchins that feed on kelp.

Several years ago, while working for Earthwatch Institute, I traveled to Prince of Wales Island in Southeast Alaska to report on a research study looking at the impact of sea otters on seagrass meadows. Ginny Eckert, a marine ecologist and fisheries scientist at the University of Alaska Fairbanks, was leading the study, and had been monitoring the relationship between sea otters and their habitat for more than a decade. She was well-versed on Estes's sea otter research, as he had been her PhD advisor.

Ginny and her research team had found that sea otters played another ecological role that could support fisheries. While the otters were consuming large quantities of shellfish, the decline of the population of Dungeness crab was leading to a rise in insect-like amphipods and isopods that feed on algae, which in turn competes with seagrass. In other words, more otters meant more seagrass. Seagrass meadows serve as a nursery for juvenile fish like salmon, herring, and rockfish. Not only that, seagrass is a carbon sink. So, by positively impacting seagrass meadows, otters may also be playing a role in carbon sequestration, which has important implications for climate change.

It was fascinating to think about the diverse ways in which individual species can impact their ecosystems. But when it came to seals, I found that we know frustratingly little about their ecological role and functions. I imagined that part of the reason for the dearth of data on seal impacts was because of how challenging they are to study. There's only so much we can understand about a species that forages underwater in areas that are difficult for humans to access. And if my January visit to Seal Island taught me anything, it was that even when seals come ashore, there are significant challenges involved in studying them. Yet without this data, how could we hope to answer questions about whether there are "too many" seals, or understand how they might be impacting fish populations—or shark populations, for that matter?

THE MALE GRAY SEALS continued to eye us reproachfully as we stood on the deck of the boat. Every so often, they slapped the surface of the water, presumably warning us to stay away from their blubber babes.

After hours rolling on rocky seas, my stomach as sloshy as the swells, we began our journey home. I knew I was lucky to have observed the breeding colony for myself, but part of me wished the seals had chosen a *slightly* more accessible location. I imagined there was so much happening on the island I wasn't able to see or experience while peering through my binoculars on the wobbly deck of the boat.

Still, I was grateful to have a visual of a gray seal breeding colony. It wouldn't be long before the white-coated pups on the island would take to the sea. After weaning from their mothers, the pups would spend a few weeks on the island before instinct and hunger compelled them into the water for the very first time. There, pups must learn how to forage on their own to survive. Some have more luck with this than others, and many seals don't survive their first year.

Not long after my journey to Seal Island, a young gray seal, newly weaned from his mother, plunged into the frigid salt water to begin his own adventurous journey. Although I can't be sure, it's possible this particular seal was born on Seal Island—it's even possible he was there when I visited. Either way, it wouldn't be long before I'd meet the young traveler in a very different setting.

AS SUNDAY TICKED INTO Monday on a cold January evening, several inches of snow had already blanketed the Maine coastline, with the promise of another foot on the way. Scott Smart, parks foreman for Cape Elizabeth, a seaside town just south of Portland, was midway through his overnight plow shift, clearing the roads as best he could before the morning commute. Around one a.m., as he was working his way through the Oakhurst Road neighborhood, he noticed a dark object on the side of the road ahead and slowed his truck. He assumed it must be a piece of turf, until it began to move. The mysterious creature scooted across the road in front of the plow like a caterpillar, shifting its weight from back to front as it shuffled toward a snowdrift. For Smart, that belly shuffle was unmistakable—the undulating creature was a seal. He radioed his supervisor, who alerted the Cape Elizabeth Police Department.

It was an unusual call for police dispatch to receive in the middle of a snowstorm. Oakhurst Road was at least a quarter of a mile from the water's edge, and the seal would have had to cross several roads and pass through various homeowners' backyards to reach it. An officer arrived at the scene shortly after. He managed to capture and lift the seal into his vehicle. He then called the 24-hour reporting hotline for Marine Mammals of Maine.

Dominique Walk had just returned home from her evening shift at a nearby animal hospital, where she works part-time as a veterinary technician, when she received the call. By day, Dominique is the assistant stranding coordinator for Marine Mammals of Maine, and she helps to operate the organization's marine mammal reporting hotline. While the group receives calls about stranded marine mammals during all times of year—not just seals, but whales, dolphins, porpoises, and even the occasional sea turtle or sea bird—this was the first call she'd received about a stranded seal in the middle of a blizzard at two o'clock in the morning.

Given the rough conditions of the roads—her truck had slid and skidded on ice for most of her drive home from the vet clinic—she wasn't able to travel to Cape Elizabeth to assess the seal herself, and she was reluctant to call upon one of the local volunteers until the worst of the storm had passed. But with an officer already on the scene, Dominique was able to guide him through the process of assessing the seal's health and condition over the phone. As soon as she heard the seal vocalize in the background, with its signature haunting wail, she determined the species—a gray seal—and the photos from the officer confirmed her assessment. Its size and features indicated it was a pup that had likely been newly weaned. Since the seal appeared to be in stable condition, she advised the officer to transport it to nearby Fort Williams Park and release it on the beach.

Fort Williams abuts Maine's most famous lighthouse, Portland Head Light, a picturesque attraction and popular tourist destination. But in the wee hours of that January morning, as the snow piled up around them, the officer and the seal were the only mammals in sight. Dominique had named the seal "Number Six," as it was the sixth stranded seal that had been reported to Marine Mammals of Maine that year. The officer successfully released Number Six on the beach near the park, placing it just next to the water. *Mission accomplished*, he thought. Number Six disagreed.

Hours later, around seven a.m., the officers learned that Number Six was traveling down Shore Road, not far from the park, which they jokingly attributed to the scent of fresh donuts at a nearby Cape Elizabeth bakery. The officers again called Marine Mammals of Maine. This time, the stranding team dispatched a local volunteer to help the officer collect the seal and bring him back to Fort Williams Park. But again, Number Six had other plans.

An hour later, the seal was spotted exploring the park, nowhere near the beach where he had been delivered. But in the daylight hours, the restless sea dog was now at risk of running into humans visiting the park and, of greater concern, their unleashed land dogs. By this time, the snow had abated, allowing staffers to collect Number Six, load him into the back of the Seal Mobile, and drive him to the rehab center in Brunswick.

Amidst his wanderings, the young seal had captured quite a few human hearts, not to mention national headlines. News stations across the country picked up the story about the determined seal that refused to return to the ocean. Alongside each story was a photograph of a snow-speckled Number Six, his big, dark eyes staring into the camera, with the blue and red flashing lights of a squad car in the background, just your standard roadside arrest of an aquatic jaywalker.

Even U.S. Senator Susan Collins shared the news about the "curious and friendly pup" in a social media post, giving the Cape Elizabeth Police Department her "seal of approval" for their role in helping the animal. But Number Six's story was only just beginning.

The wayward traveler spent the first couple of days at the rehab center sleeping and suckling on his flippers, curled up beside a seal stuffy for comfort. He was exhausted.

One morning not long after his arrival, I decided to pay the team a visit. We were in the midst of another, albeit far milder, snowstorm, and the roads were empty as I cautiously drove the short distance to Brunswick. Thick, fluffy flakes bounced off my windshield.

By the time I arrived at the center, Lynda, Dominique, Katie, and Lexi Wright, the team's community engagement coordinator, were already hard at work. In addition to Number Six, there were three other gray seal pups receiving care—two additional strandings from Maine, and one that had been transported north from Cape Cod in need of significant rehab support. Lynda was about to join a conference call with NOAA Fisheries, but suggested I tag along with Lexi to observe her seal care rounds. "Just beware of Number Six," she said. "He's had the *stinkiest* poops all morning."

"I've never seen seal poop before," I replied.

"Well, get ready. His are real blowouts."

I followed Lexi through the supply rooms connecting the office to the animal care center. On the other side of the door to the center, a set of stairs descended into a large, warehouse-style room filled with pools, tanks, and gated animal enclosures. As we entered the room, a loud and low howl, reminiscent of the spooky sounds emanating off of Seal Island, emerged from one of the two large pools across the room. "Well, don't you have a lot to say this morning," said Lexi. Turning to me, she added, "That would be Number Six." It appeared he'd already gained a reputation as the most vocal of the current seal patients, as he didn't hesitate to let the staff know when he was in the mood for fish . . . which was always.

Number Six had been paired with another gray seal, Number Four, which had also stranded in Cape Elizabeth in January. While it wasn't yet feeding time for the seals, Six's howls—and Four's efforts to hurl herself in and out of the pool on repeat until she'd flooded the deck—suggested they disagreed. Meanwhile, a third gray seal, Number Two, quietly minded his business in the adjacent pool. Given his weakened condition—a boat collision had left him with a fractured flipper—he wasn't yet ready for a pool-mate.

The layout of the rehab center was a far cry from the captive seal habitat at the New England Aquarium, which made sense. The aquarium's goal was to create a safe and enriching space for harbor seals to live out the entirety of their lives, facilitating bonds between the caretakers and the animals. The goal at the rehab center, on the other hand, was to improve an animal's health as quickly as possible so it could be safely released back into the wild. It was effectively a hospital for seals. The staff maintained as much distance from the animals as they could, limiting their interactions to feedings, cleanings, and medical care. For seals well enough to be released back into the wild, their ultimate survival depended upon their having as little trust and interest in humans as possible.

To curtail the seals' innate curiosity, each of the large pools had privacy fences that rose above the decks to prevent the animals from being able to watch humans milling about the room. The staff hung towels around the fenced animal enclosures for the same reason. But that morning, two of the towels in one of the enclosures had separated slightly to create a small opening. As I walked by the enclosure, a single eyeball, which belonged to the Cape Cod seal, peered out. The seal still had patches of lanugo he had yet to molt, meaning he had likely stranded before he had been fully weaned. He was probably just a few weeks old.

Meanwhile, Lexi had gathered the supplies she needed and was crouching over what appeared to be a large, flat scale positioned beside one of the tan kennels. "So I'll grab him, lift his blanket, and try to put him in here," she said. "I'll need your help pushing the kennel up, unless he's too flaily. We'll just try to make this as quick and painless as possible."

"That sounds great," I said, despite having absolutely no idea what we were doing, nor what she meant by a "flaily" seal. Lexi cautiously opened the gate to the pup's enclosure, at which point, mouth open and teeth bared, he screeched in fury and immediately tried to hurl himself out of the enclosure. She blocked his exit with her body, threw the towel over him, covering his eyes, and quickly wrapped her arms underneath him to lift him out.

As I stood staring open-mouthed at Lexi and the seal, shocked by the wrath of this tiny, adorable animal, I realized my time to help had arrived. As instructed, I lifted the open end of the kennel toward Lexi, who managed to lower the snarling seal, still covered in his towel, inside the crate, gently tucking in his flaily flippers as he attempted to slap her with them. "This is so easy with two people," she said. I wondered which part of this process she considered to be easy. We then carried the kennel onto the scale, and Lexi recorded the weight measurement. After she returned the furious seal to his enclosure, she recorded the weight of the empty kennel and added the final difference to the whiteboard beside his enclosure: 15.7 kilograms, about 34 pounds.

"Now comes the fun part," she said, as she picked up an empty plastic kiddie pool and headed toward the back entrance of the center. "Could you grab that shovel?" Outside, the snow was steadily falling, the soft powder blanketing the trees. "Since there's two of us, we can really load it in," said Lexi. She ran back to grab a second shovel while I started filling up the pool with snow. When it was nearly overflowing, we carried it back inside.

Lexi unlatched the door of the seal's enclosure and we lifted in the snow-filled pool. The pup immediately scrambled on top of the snow pile, burying his face in the frozen powder while flinging snow with his flippers. Since the Cape Cod seal was still too young to have full-time water access, the snow would help him stay cool while also serving as a form of enrichment.

Before returning to the office, we stopped in the adjoining admit room. One side of the space was set up for seal triage, with an examination table, an IV stand, several nebulizer machines, and various other equipment. The other side of the room was reserved for food prep and data collection, and contained large freezers filled with fish, a sink, blenders, tubes, funnels, and a whole host of other supplies. A long table was set up on one side of the room with four clipboards, one for each seal patient. Each held a thick stack of forms. On the wall was a massive whiteboard with information about each seal, including their stranding location, date of admission, weight at arrival, current weight, and feeding protocols.

I picked up Number Six's clipboard and skimmed through the observation notes written up in the days after his arrival—"Squirmy . . . Sleeping with some twitching, maybe dreaming . . . Swallowed fish well . . . Suckling on stuffy . . . Someone in pool defecated small amount of feces." Lexi said that the summer interns were renowned for their colorful descriptions of seal poop, her favorite being "golden nuggets."

Lexi was in her early twenties and had first started working with the organization as an intern a few years back. She had studied marine biology in college but hadn't planned to work with seals specifically. "But I quickly learned there's nothing else I want to do ever," she said. "They're just so charismatic." But it wasn't just the seals that drew Lexi to the organization. "A supportive and strong female-led team is something you don't come by every day in this field," she said. After her internship, she continued volunteering at the center until Lynda was able to raise enough funds to hire her full-time.

Back upstairs in the office, Lynda had finished her call. After she and Lexi debriefed on the seal patient status and feeding strategies for the day, she sat with me on the couches for a chat. I asked her whether seals ever ate lobster in addition to fish. While I'd read and heard quite a bit about the frustrations of fishermen, particularly gillnetters in Cape Cod dealing with gray seals, I hadn't heard much about the impacts of seals on Maine's highest-value ocean resource. "Most diet studies don't show much in terms of seals eating crustaceans," she said. "They mainly stick to fish." She explained that while seals occasionally mess around with traps as they try to access the fish bait used to lure in lobsters and crabs, they generally don't want to work that hard for a meal.

"But that doesn't mean lobstermen love seals," she added, a subject she was particularly well-versed on. Not only is Lynda's partner a lobsterman, but she has a commercial lobster license herself.

Back in 2011, when Lynda founded Marine Mammals of Maine, she was also lobstering and working various other part-time jobs to make ends meet. Initially a one-woman show, she often had to answer any calls about stranded seals on her own, even if they came in over the reporting hotline during her long days on the lobster boat. Her captain, who wasn't a fan of seals, grumbled about her split focus, but he begrudgingly allowed it. I can't imagine anyone has ever questioned Lynda's work ethic.

But a few years later, dividing her time between running her nonprofit, lobstering, and managing various other jobs to make ends meet became overwhelming. She was exhausted. She decided to invest everything she had, financially and otherwise, into Marine Mammals of Maine. Outside of her family and friends, no one seemed to have faith she could hack it running a business, she told me. While her former colleagues and supervisors acknowledged she'd always excelled at animal care, they doubted her ability to fundraise, manage business operations, and make tough decisions.

"But you know what I said to them?" she asked, her eyes twinkling. "Hold my beer. Watch me."

Now, more than ten years later, Lynda has grown the organization from a one-woman show to managing six staff, coordinating a statewide network of thousands of passionate volunteers, and supporting population research efforts alongside marine mammal rescue and seal rehab. The growth in Lynda's operations and capacity seems to have coincided with the growth in Maine's seal numbers, and with the increasing frequency of human-seal interactions.

But something confused me about the recovery of seal populations. With some rebounding species—sea otters and wolves, for example—wildlife officials reintroduced small numbers of animals by translocating them from other regions to kick off the recovery process. But that wasn't the case with seals. Seals had recolonized their historic habitat without any direct intervention.

So then, where did they all come from?

IN 1999, TWENTY-THREE-YEAR-OLD OWEN Nichols was leading a nature tour in Cape Cod when he happened across a dead gray seal with an unusual, human-made mark on its back. The seal had been branded with the tag "B-563." Owen reached out to a local naturalist, Valerie Rough, who was one of the first people to study the return of seals on Cape Cod. Rough recognized the tag immediately. She connected Owen to a Canadian scientist who studied gray seals on a remote island off the coast of Nova Scotia.

The scientist shared with Owen the information he had about

B-563. The seal was twenty-six years old and for many years, he'd been a dominant bull on Sable Island, the breeding grounds where he'd first been tagged. But then he'd disappeared. This was the first reported sighting of him in years.

While an estimated 800 gray seal pups are born each year on Seal Island, Maine, where I'd spent a frigid January morning, one *hundred* times that number—or roughly 80,000 gray seals—are born on Sable Island, according to 2021 estimates. I needed to learn more about the seals on this tiny island, a treeless spit of sand in the North Atlantic, and their unique relevance to the New England seal story.

6

The Graveyard of the Atlantic

A TWO-SEATER GUMDROP OF a helicopter dips low over the beach on a remote island in Canada, nearly two hundred miles off the southeast coast of Halifax, Nova Scotia. Startled by the roaring aircraft, dozens of gray seals wriggle from the sand into the surf. Heads bobbing in the waves, their large eyes stare up at the mysterious creature passing overhead. The chopper banks to the right, over the high dunes, as it overtakes a band of wild horses galloping frantically beneath the deafening whir of the propeller. At last, it touches down on a patch of sand surrounded by dune grass, resting atop its inflatable pontoon-fitted skids. Emerging from the helicopter, Jacques Cousteau grips his signature red beanie, preventing it from being carried away by the relentless winds, as he steps onto Sable Island.

So begins the closing scene of Cousteau's 1982 documentary, *St. Lawrence: Stairway to the Sea*. The film features the legendary oceanographer retracing the path carved four hundred years earlier by French explorer Jacques Cartier along the St. Lawrence River. In the final few minutes of the documentary, Cousteau and his crew cross the Gulf of St. Lawrence into the Atlantic Ocean aboard the research vessel *Calypso*, anchoring just off the coast of Sable Island. Cousteau narrates the scene, describing the island as

". . . a shrinking dune, pounded and constantly reshaped by the surrounding seas. A small, lonely world of seals and wild horses and abandoned dwellings, which a handful of volunteers is trying to save from despoliation and oblivion . . ."

Donning tall black rubber boots, a fur-lined winter coat, and his signature cap, Cousteau is greeted by Zoe Lucas, a young Canadian naturalist who has been living on the island. Zoe guides Cousteau along the sandy beaches, describing some of her dune restoration efforts to prevent beach erosion. "Tell me," says Cousteau, smiling down at her. "Do you love the island?" "Certainly," she replies. "It's home, I suppose."

A SMALL, CRESCENT-SHAPED SANDBAR teetering on the edge of the continental shelf, Sable Island is a Cheshire Cat–like smile in the sea, less than thirty miles long and no more than a mile wide. The island sits at the exact meeting point where the cold Labrador Current collides with the warmer waters of the Gulf Stream, setting the stage for intense and frequent storm systems. Sable Island is one of the most hurricane-prone regions in Canada, and during the summer months, it's considered to be the foggiest place on earth.

For centuries, the island was notorious for luring even the most skilled seafarers to their doom. Its long, treacherous shoals extend like tentacles to wreck any vessel that ventures too close to its shores. Sable is referred to as the "Graveyard of the Atlantic" — over three hundred shipwrecks have been recorded there.

While few humans are aware of the island's existence, those who are often develop a cultlike fascination with its natural wonders and murky human history. The remote landscape is a haven for wildlife. Sable is a key stopover point for hundreds of species of migratory birds, with more than a dozen nesting on the island. It's also home to a variety of endemic plants and invertebrates, along with one of the most extensive dune systems on earth. But the island's most iconic residents are its wild horses—roughly five hundred descendants of animals believed to have been abandoned on Sable in the 1700s. The horses, which are actually feral rather than wild, have adapted to the island's harsh conditions and unique ecosystem, supplementing their beach grass diet with kelp and seaweed that wash up on the shores.

Sable Island is also a particularly popular destination for seals. Harp, hooded, and ringed seals are occasional visitors, journeying south off the pack ice in northern Canada and the Arctic, and the island is home to a small resident population of a few hundred harbor seals. But the most notable seal inhabitants are the nearly 400,000 gray seals that haul out on Sable's shores. The island is home to the largest breeding colony of gray seals in the world.

One species that has consistently struggled to survive on Sable, however, is humans, despite nearly five hundred years of settlement attempts. These days, the only human residents on the island are a small group of seasonal researchers, a couple of government staff caretakers, and one intrepid and committed naturalist named Zoe Lucas.

An artist by training, Zoe traveled to Sable Island for the first time in 1971, when she was in her early twenties. She was immediately captivated by the landscape. "It was like being in a watercolor painting," she told me during a surprisingly clear video call from Sable, powered by the island's new Starlink satellite internet. "The incredible breadth and openness, the quality of light on the island . . . the colors, the sensations, everything."

Her first visit lasted just a few days, but she was determined to find a way to return. In 1974, while a student at the Nova Scotia College of Art and Design, she signed up to volunteer as a cook and, eventually, a field technician for the seal research team at Dalhousie University, studying harbor seal behavior on Sable

Island. By the early 1980s, she was working full-time on the island and involved in a wide range of research and monitoring projects. Since then, she's worked on extensive vegetation and horse surveys, along with projects focused on marine debris, shark predation, seabirds, cetacean strandings, and more. For decades, Zoe was one of the only people living year-round on Sable. Her devotion to the island, keen observations, and meticulous recordkeeping have inspired comparisons to Jane Goodall and Dian Fossey.

Over the past fifty years or so, Zoe has observed significant changes in the populations of the island's resident seals. In the 1970s, there were thousands of harbor seals living and breeding on Sable, but by the 1990s, the population began to decline significantly. According to scientists, part of the reason for this decline might relate to the explosive population growth of gray seals. The number of gray seals on Sable Island has risen from less than 3,000 in the 1960s to nearly 400,000 today, and it's possible gray seals are simply outcompeting the smaller species.

When Zoe first arrived on Sable, gray seals gathered mostly on the eastern tip of the island, she said, but these days, particularly during the winter pupping season, it's impossible to go anywhere on the island without running into gray seals. "Sometimes you look into the water and there are thousands of ocean eyes looking back at you." She's even had a gray seal block the door to her field cottage. "I had to climb out a window to get out."

Today, Zoe serves as the president of the Sable Island Institute, delivering presentations to packed auditoriums in Nova Scotia about life on the shape-shifting North Atlantic sandbar. Part of the allure of the island is that it remains largely out of reach for humans.

THE MORE I LEARNED about Sable Island, the more fascinated—some might call it obsessed—I became. I dove into research articles, books, and documentaries featuring the island's rich natural history, but I kept returning to the same question: how can I experience this enigmatic place for myself? Part of my intrigue stemmed from learning about the critical role Sable Island's seals have played in New England's seal story.

Like so many others who have become captivated by Sable, my desire to visit was repeatedly stymied by the logistics of getting there. While there were some limited opportunities for members of the public to visit the island during the summer months, I specifically wanted to go in the dead of the winter, during gray seal pupping season.

Unfortunately, the only people who travel to Sable Island in the winter are government officials and seal scientists, in part because of the added safety and logistical challenges associated with Nova Scotia's unpredictable winter weather. By mid-January, more than halfway through the pupping season, I was no closer to my goal of reaching the island. Still, I was thankful to have observed firsthand the gray seal colony off the coast of Maine. I focused on learning as much as I could, even if from afar, about the connections between the Sable Island seals and the recovery of the U.S. populations.

In the 1990s, as the small breeding colonies in Massachusetts and Maine began to grow, scientists observed additional seals with brands and tags from Sable Island. They wanted to find out whether there were genetic differences between the U.S. and Canadian populations. Were the gray seals in the U.S. a distinct group, perhaps descendants of the few seals that managed to survive the bounty hunts, or were they immigrants from the Canadian colonies? To answer this question, Stephanie Wood, a biologist at the University of Massachusetts Boston, and her research team collected tissue samples from seals in different breeding colonies in the U.S. and Canada. Their findings revealed that there were no significant genetic differences between *any* of the populations. The U.S. gray seals weren't a distinct population, but a genetically representative sample of the entire Northwest Atlantic gray seal population.

In other words, without the thriving population of gray seals on Sable Island, it's unlikely that the U.S. colonies would have ever recovered. Not only that, the connection between these colonies raises important questions about the effectiveness of a U.S. seal cull. The Sable Island seal population accounts for roughly 80 percent of the gray seals born each year in the Northwest Atlantic. If U.S. seals were culled, what would prevent seals from Sable Island from traveling south to fill in the gaps? Wildlife doesn't exactly adhere to human-imposed borders.

NELL DEN HEYER IS an ecologist with Canada's Department of Fisheries and Oceans, or DFO, and has spent fifteen years studying gray seals on Sable Island. "I'm there for the science, but the experience of working on Sable, it's just amazing how different it is every year," Nell told me on a video call months earlier. "It's an exciting place to go." Nell and her research team have been studying the population size and growth rate of the island's seals, where and how they're foraging, and what they're consuming. "Seals like to breed in remote places," said Nell. "While it's hard to get to Sable Island, once we get there, we can be right there in the colony working."

The winter pupping season offers the perfect opportunity to study the population, but six weeks isn't a lot of time when it comes to monitoring hundreds of thousands of seals. In addition to counting individual pups, the team has deployed a variety of satellite and acoustic tracking devices to better understand seals' feeding behaviors. They even attach small cameras to some of the animals, which provide a unique seal's-eye view of the ocean.

According to Nell, the satellite tracking data collected from seals tagged on Sable Island have shown that individual animals don't necessarily stay in or around the island after the breeding season. The seals will forage anywhere from Labrador to Cape Cod.

BEFORE THE INVENTION OF radar technology, Sable Island was the bane of every sea captain. The biggest challenge was its precarious position in the center of a major shipping lane that connected Europe to the Americas. By the end of the eighteenth century, driven by economic and humanitarian concerns, the Canadian government commissioned the development of a lifesaving station to rescue anyone shipwrecked on the island's shores, as well as their cargo. In 1801, James Rainstorpe Morris, a former Navy man, was appointed the first superintendent of the Sable Island Humane Establishment.

The island's notorious weather proved to be an immediate challenge for the new settlers, and the crew struggled against wind and rain to build their shelters. During a brief reprieve between storms, Morris ventured to the western tip of Sable, where the island's most populous residents, the seals, were gathered en masse. There were ". . . *about 200 on the point of the bar, others off and on playing,*" he later wrote in his diary. To get a sense of their movements and behavior, Morris ran straight into the middle of a herd of eighty seals. "*They were in general large & moved very slow over the sand,*" he wrote, "*they ran off from me in all directions, and Judged that with ease, five men would if well arranged, have killed 40 of them, in 5 minutes.*" The good news, for the seals at least, was that Morris and his family didn't have much of an appetite for seal meat, nor did his crew. But they did occasionally kill seals for their oil, which was used for lighting and cooking.

Late one January evening in 1803, Morris was alerted to a series of alarming cries, like a human in distress, heard north of his house. A winter storm had recently swept across the remote island, leaving several inches of snow in its wake. Bracing himself against the wind, Morris set out to search for the source of the *"horrid noise."* As it turned out, the cries weren't those of a distressed human, but rather, a distressed seal. Morris later wrote that the gray seal pup *". . . appeared to be about 10 days old, got astray from the Mother . . . the track was found near one mile from the beach to the house . . ."*

THE HUMANLIKE CALLS AND cries of seals have inspired myths and legends that date back to antiquity. In 1939, British ecologist Frank Fraser Darling wrote, "There is no creature born, even among the greater apes, which more resembles a human baby in its ways and its cries than a baby grey seal." In Greek mythology, the captivating voices of sirens—mermaid-like sea nymphs known for their nefarious tendencies to lure sailors to their deaths with their beautiful songs—may have been inspired by the haunting vocalizations of seals.

In Homer's *The Odyssey*, during his journey home from the Trojan War, Odysseus was forced to sail his crew past the dangerous island of the sirens. To protect himself and his shipmates, Odysseus commanded the crew to plug their ears with softened beeswax, then tie him securely to the ship's mast. As they rowed past the island, he desperately struggled against the ropes that held him, enchanted by the sirens' calls, but his crew rowed on, heeding his previous orders.

In the early 2000s, Karl-Heinz Frommolt, a German bioacoustics scientist, tested the seal versus siren theory. Frommolt set out on an archaeo-acoustical expedition to the Galli Islands, off Italy's Amalfi Coast, long believed to be the real-life location depicted in Homer's epic tale. Frommolt's team positioned speakers on a set of rocks known to amplify sounds from the island. The team broadcasted the voices of two German opera singers at various pitches, followed by recordings of monk seals, which are native to the region. He later reported that the cries of monk seals were far louder than those of humans and could indeed be heard offshore from the direction in which Odysseus and his crew would have been sailing.

I could only imagine a sailor's experience traveling past Sable Island hundreds of years ago—to hear the haunting moans of thousands of seals emanating from an island shrouded in mist and fog. Mythical sirens luring sailors to their demise might have been a more plausible explanation for the noise than a discordant chorus of seals. (Matt Groening, creator of *The Simpsons*, took this notion a step further during an episode of his animated show *Disenchantment*, portraying sirens as a herd of amorous walruses eager to mate with passing seafarers.)

Of course, the gray seals on Sable Island do much more than vocalize (and potentially lure ships to their doom). The seals are intricately linked to their environment, in ways we're only just beginning to understand. Much like the arrival of bald eagles on Seal Island in Maine, black-backed gulls arrive in droves on Sable in the winter months to align with the seals' pupping season. "There are placentas and dead seals everywhere," said Greg Stroud, a Parks Canada operations coordinator for Sable Island and an avid birder. "It's the only reason those gulls come here, because of the gray seals."

But one of the most astounding examples of seals' ecological impacts on Sable came to light in recent years. For centuries, Sable Island's wild horse population had fluctuated between 200 and 400 animals, but in recent decades, the population began to increase significantly, between 450 and 550 animals. Meanwhile, Zoe had identified an interesting pattern related to the growth of beach grass. While conducting vegetation surveys across the island, she discovered that "where the gray seals had been breeding, the beach grass was very different. It was more robust, deeper in color, bigger leaves, taller leaves, denser."

By that time, the population of seals on Sable Island had exploded in number, from less than a few thousand animals in the 1960s to roughly 400,000 by 2014. Increasing seal numbers meant a dramatic increase in the amount of nutrients that were being transferred from the sea to the land. The seals were effectively fertilizing the windswept grasses with their waste, placentas, and carcasses.

In 2016, a team of biologists at the University of Saskatchewan, led by Philip McLoughlin and Keith Hobson, determined that the horses had changed their foraging habits. They had begun to gravitate toward the dense grasslands where the seals congregated. While the biologists cautioned that the findings did not yet reveal a direct link between the growth in the seal and horse populations, they reveal intricate ecological connections between these two species.

And it wasn't just the beach grass the seals had been fertilizing. In 2021, Nell and her research team at DFO published an analysis of 20 years of satellite data that provided visible evidence of the extent of seals' nutrient transfer capabilities. During the late fall and early winter, as gray seals hauled out in massive numbers on the island, the nitrogen emanating from their poop appeared to be creating a surge in phytoplankton in the waters surrounding the island that far exceeded levels in comparable regions. Phytoplankton are tiny, plantlike organisms that float near the surface of the ocean and feed a wealth of marine species, including other plankton and fish. Phytoplankton also produce oxygen and play a role in the carbon cycle. Typically, phytoplankton bloom in the fall, then decrease in biomass during the winter. But the satellite images revealed swirling cyan plumes of chlorophyll, the surface-level signature of phytoplankton, during these ordinarily unproductive months. As Nell and her team demonstrated, the ecological impacts of the seals on Sable Island can quite literally be seen from space.

JOE ROMAN IS A conservation biologist and author of the book *Eat, Poop, Die*, which explores the diverse ecological functions of wildlife (and boasts one of the best book titles out there). I spoke to Joe on a call, and he explained that similar impacts have been observed on Surtsey, an uninhabited volcanic island off the southern coast of Iceland. The arrival of gray seals in the 1980s created a steady nutrient supply, which enriched the vegetation along the lower shores.

Because of the isolated, uninhabited nature of islands like Sable and Surtsey, scientists have been able to study complex ecosystem dynamics in ways that are far more challenging in more populated areas, where many other factors might be influencing the system. But I wondered how, or whether, it would be possible to tease out the specific impacts of seals—their role and functions within their environment—in a far more populated region, like New England. How could wildlife managers in the U.S. weigh the risks and benefits of culling seals, if faced with that decision, without understanding how their removal might affect other species and their habitat? I asked Joe about this. He said that case studies like Sable Island and Surtsey were useful examples of the powerful ways in which seals can shape their environment. But he added that it's also helpful to consider data on the ecological role of other marine mammals. As marine mammal populations recover in regions around the world, scientists are beginning to witness fascinating behavioral changes as well.

In Australia, for example, after commercial whaling nearly killed them off, the population of humpback whales has rebounded thanks to dedicated conservation efforts. But as the humpbacks returned, killer whales began to hunt their calves. To protect their offspring, humpbacks began to fight back against the killer whales in ways that hadn't previously been observed. Even more surprising was that they didn't just fight killer whales to protect their own species; they also did so to protect other species, including seals.

"These are things that we don't tend to see when animals are rare," said Joe. But as the population of a given species expands, ecological and behavioral changes become more evident. "That, to me, is the coolness factor. It's fun to think about how animals might have impacts beyond what we typically would think."

AFTER MONTHS OF EMAILS, phone calls, and what I hope was very polite harassment of Canadian officials and scientists, I'd begrudgingly accepted that a Sable Island journey wasn't in the cards for me. But on a cold January morning, I received word that a single seat had become available on a government-chartered flight, departing within a week. If I could get myself to Halifax in time, they said, the seat was mine.

Days later, brimming with excitement, I gripped the armrests aboard my Air Canada flight as we descended through the fog before touching down in Halifax. The customs line moved quickly, and I was soon being waved forward by a hulking Canadian border official. "What's the nature of your visit?" he asked gruffly, as he studied my passport. I attempted to suppress the bizarre smile that had been plastered to my face since we'd landed. "Pleasure!" I said, too loudly. The agent looked up, his eyes narrowing as he assessed my threat potential. I stared back, grinning like a lunatic. "Who are you visiting?" he asked, watching me. "Who?" "Yes. *Who.*" He glared impatiently as I debated how to answer. Technically, I was visiting nearly half a million seals, but I suspected my plans to mingle with wild creatures wouldn't ease the concerns of this imposing border agent. Worried I was teetering on the edge of detainment, I tried a different approach. "Well, I guess I'm mostly here by myself. But you could say I'm visiting a team of research biologists. Oh, and the Parks Canada staff. Maybe you know them? You see, I'm doing research, so I guess that technically qualifies as a work visit, now that I think about it. But honestly, it's just such a pleasure to be here, so that's why I said pleasure earlier." As I continued rambling, the agent aggressively scribbled on the customs form, then waved me forward with a grunt. I was in.

Unfortunately, my excitement about the following day's adventure waned as soon as I reached my hotel and caught up on my email. Sarah Medill, the Parks Canada operations coordinator on Sable Island who had arranged the flight, alerted us that a short-lived but intense storm had hit the island the prior evening, and Sable's "runway," a long stretch of hard-packed sand, was presently underwater. She would check the conditions again early the next morning and report back.

I awoke to a stunning scarlet sunrise over the Halifax harbor, but Sarah's early morning update was less welcome. The runway was still flooded, and the flight was officially canceled. Even worse, the weather models indicated a series of back-to-back storms heading toward Nova Scotia. It was unlikely they'd be able to get a flight to the island for at least several days, if not longer. I knew if I extended my stay in Halifax for another week, I'd still be at the mercy of the winter weather gods, not to mention the cost and logistical challenges involved. It was a risk, but I wasn't ready to give up.

Over the following days, nearly a foot of snow blanketed the city of Halifax. I watched the large, fluffy flakes float past the floor-to-ceiling windows in the city's public library, where I spent hours in the reference room reading every book about Sable Island

I could find. I wandered the quiet streets, exploring cafés, shops, museums, and restaurants. Everywhere I went, I asked people what they knew about Sable Island, but was surprised by how few people had even heard of it. Among those who had, most knew only about the island's "ponies." No one mentioned anything about the seals.

By Saturday, the snow had ceded to heavy rains. Despite the dreary weather, Sarah's update was a welcome one—the latest aviation forecast predicted clear skies for a brief twelve-hour window the following day. While there still wasn't enough dry beach to safely land a plane, Parks Canada was eager to swap field staff and deliver supplies to the island after a week of grounded flights, so they'd chartered a helicopter. I'd no longer be able to stay overnight on the island, but I'd at least be able to experience it for the day.

Elated, I shared the news on my family text chain. My mom was the first to reply, responding in her signature emoji language: a smiley face, a helicopter, and a DNA double helix.

Me: *Why the double helix?*

Mom: *That's a double helix?! I thought it was helicopter goggles!*

The following morning, with my backpack packed and ready to go, I held my breath as I checked the latest update from Sarah: *Just confirming that everything is a GO. Have a great day and those of you coming here, see you soon.*

7

The Corkscrew Seal Mystery

IN JANUARY OF 1993, while conducting a survey of Sable Island's beaches, Zoe Lucas encountered something strange: a harbor seal carcass with a gruesome, oddly shaped wound. Finding a dead seal with visible injuries wasn't out of the ordinary—sharp-toothed predators like sharks and killer whales were known to prey on seals in the waters around Sable—but it was the unique shape that surprised her. The laceration was long, deep, and clean-edged, but most notably, it spiraled around the entirety of the seal's body, much like a corkscrew.

By the end of that year, Zoe had documented nearly one hundred cases of these corkscrew wounds on both gray and harbor seals, mostly pups that had been weaned from their mothers. She shared her findings with various scientists, some of whom theorized it was the work of great white sharks. "But I couldn't understand how a white shark could inflict those wounds," Zoe told me on a call. Most of the carcasses were found in the winter months, long after white sharks migrated south to warmer waters off Florida and the Carolinas. Not only that, the shape of the wounds bore little resemblance to typical white shark predation attempts, which tend to be slash cuts, punctures, and rake marks consistent with a single bite.

Meanwhile, more and more carcasses were washing ashore. In 1996, Zoe recorded over four hundred seals with corkscrew wounds. A year later, there were six hundred. She began writing letters to shark scientists around the world, from South Africa to Australia, and across North America, describing her findings. In each envelope, she enclosed photographs of the corkscrew lacerations. None of the shark experts she reached out to had seen anything like it. Eventually, Zoe connected with Lisa Natanson, a shark scientist in Rhode Island who worked for the National Marine Fisheries Service. Natanson agreed to help Zoe in her quest to identify the Sable Island seal killers.

While most of the wounds observed had been on dead seals, the cuts themselves were typically quite fresh, some warm to the touch, others still bleeding when they came ashore. It was clear the attacks were occurring close to the island. Between 1993 and 2001, nearly 5,000 seal corpses with visible wounds were examined on Sable Island. While a small number of the wounds were slash marks consistent with white shark attacks, 98 percent were the spiralized lacerations. "Even when you've only got half a seal left," said Zoe, "it's easy to see the characteristics that indicate that particular wound."

The research team considered several non-shark-related possibilities, but the wounds were inconsistent with the most obvious explanations, such as boat propellers or entanglement in fishing nets or lines. Additionally, Sable's remote location and long shoals prevented boats and ships from nearing its shores. Pack ice was considered as a possible cause, but it was deemed too far from Sable, even during the winter months.

They'd ruled out white sharks as a culprit for the corkscrew wounds, but there were several other shark species to consider. On rare occasions, tiger sharks had been reported in Sable's waters, but they were only seen as juveniles, an age group that doesn't feed on marine mammals. Porbeagles similarly had never been observed feeding on marine mammals. Other species, like shortfin makos and blue sharks, were also seasonal, and were absent during the months when most of the gray seals were attacked. That left just one surprising candidate—Greenland sharks.

Greenland sharks are a slow-moving, enigmatic species that can live for over four hundred years—the longest-living vertebrate known to science. They're thought to reside mostly in the cold, dark waters of the Arctic and North Atlantic, but have been spotted in waters well outside of this region, as far south as the coast of Belize. As predators, Greenland sharks are generalists and scavengers, known to feed on a variety of fish, seabirds, squid, crabs, and even an entire reindeer, including its antlers (it's thought the shark scavenged a drowned reindeer). They are also known to feed on seals—harbor, ringed, harp, and hooded seals have been found in their stomachs in different regions—but it's generally unclear whether they were actively hunting the seals on those occasions, or simply scavenging their remains.

Greenland sharks have a distinct biological characteristic that made them a noteworthy candidate for Zoe and Natanson—the shape of their teeth. The researchers hypothesized that the sharks used their pointy, needlelike top teeth to grip the seals and hold them in place, while their lower teeth, which are shaped more like a blade, sawed through the tissue. This could have created a tear along the collagen fibers that wind diagonally around the seal's body, explaining the corkscrew pattern.

In 2010, Zoe and Natanson published their analysis in the *Proceedings of the Nova Scotian Institute of Science*. The findings were widely shared by media outlets around the world, inspiring multiple documentaries.

For some, the mystery had finally been solved, but others remained skeptical. Why would a slow-moving shark expend so much energy attacking a seal without even feeding on the carcass? And why now?

It would be several years before a shocking discovery on the other side of the North Atlantic unveiled new evidence about the Sable Island seal killers.

THE SUN WAS STREAMING through the large windows of the overheated airport hangar as I slowly worked my way into my survival flight suit. The stiff, rubber garment would have made me feel like I was in *Top Gun* if it didn't force me to waddle like a penguin. Beads of sweat gathered along the backs of my knees as I struggled with the zipper. Per the orders of the Canadian government, the suits were required attire for anyone flying over a mile offshore by helicopter where the ocean temperature was below fifty degrees Fahrenheit. Since we'd be traveling nearly two *hundred* miles over an ocean whose January temperature averaged in the thirties, I wanted to be extra careful with that zipper.

As I contemplated the various scenarios in which I might need a survival flight suit, prompting even more sweat, Greg Stroud, the Parks Canada staffer I'd been chatting with that morning, brought me back to the moment. "You see," he said, "consensual sex is not a thing for gray seals." Greg had arrived at the hangar to deliver supplies for staff on the island. He'd remain on the mainland for another two weeks before heading back out to Sable to swap places with Sarah, his Parks Canada counterpart. He'd just finished describing an alarming-sounding scene he'd witnessed on Sable where a small group of female seals attempting to return to the ocean to forage were forced to contend with roughly thirty males blocking their entrance in the surf, eager to mate.

Pups will be everywhere at this time, he continued, but warned that a good number won't survive long enough to make it off the island. "It can be quite distressing for tourists," he said. When they encounter a dead or dying seal pup, they're often confused why the staff don't intervene to help the seals. But because the populations are wild, human interference is forbidden. "There's a thin line between life and death on Sable Island," said Greg. I presumed he was referring to the wildlife, but as I looked out at the icy tarmac and considered the centuries of shipwrecks and failed settlement attempts I'd read about, I imagined the same might be true for humans. I took a deep, quiet breath.

With the luggage weighed and packed, our pilot, Dave, gave us a nod. I said goodbye to Greg and the other Parks staff who had arrived with supplies to hand off, then waddled after my fellow passengers onto the tarmac for my first-ever helicopter journey. There were seven of us on the flight: two Parks Canada staffers swapping places with their counterparts (six weeks on, six weeks off for all Sable Island staff), two videographers that would be staying on the island for a couple of weeks to produce a nature documentary, the pilot, copilot, and me. Bags and boxes filled with groceries, equipment, and other supplies were stuffed in every available nook and cranny on the aircraft and tightly secured with straps and netting.

Minutes later, we were wheeling down the tarmac, rotor blades roaring. As we lifted up, mounds of powdery snow blew across the runway, and we were soon flying over frosted trees and half-frozen lakes and ponds, headed for the coast.

I pressed my face against the window. Through the thick glass, the choppy waters of the North Atlantic resembled a Bob Ross painting—happy little whitecaps. A few of my fellow passengers' heads began to nod, but I kept my eyes glued to the waters below. I watched the helicopter's shadow as it moved along the ocean's surface and over the tops of cotton-candy clouds.

An hour later, I heard the pilot's muffled voice in my headset. As I reached to adjust the volume, the helicopter dipped slightly, and the western tip of Sable Island appeared in the window opposite mine. Instinctively, I lunged toward it, an invisible magnetic force drawing me to Sable's shores. Thankfully, my thrice-buckled safety straps spared me the embarrassment of diving across the laps of my two seatmates for a better view.

We soon crossed over the island and flew along the southern shore. Waves crashed against the long shoals. The breakers overlapped like tissue paper across the surface of the emerald water before ceding to the shore. Between the blue-green water and white sand beaches, it would have been easy to mistake Sable for a tropical paradise if it weren't for the frigid wind whipping through the cabin. Well, that and the fact that scattered across those white sand beaches, gathered in the shallow surf, and speckled throughout the high dunes and inland ponds were at first hundreds, then thousands, then, amazingly, tens of thousands of dark, glistening, blubbery specks—seals.

As we reached the eastern tip of the island, a mere twenty-six miles end to end, the helicopter banked hard to the left and we turned inland toward Sable's main station, a collection of a dozen or so buildings: living quarters, food storage sheds, and a long-standing weather station. As we began our descent, bands of wild horses, their long, crimped manes majestically backlit by the sun, abandoned their peaceful grazing along the dunes and galloped away, escaping the deafening sounds of the propeller.

As I stepped out of the helicopter, the force of the wind threatened to immediately launch me back into the air. I was soon greeted by Sarah. Dressed head to toe in pine green Parks Canada attire, which included a hat, neck covering, and gloves, Sarah's only exposed skin was a round patch of her face between her eyebrows and chin. After exchanging a few words with the pilot, she guided me toward one of the nearby buildings.

To my overwhelming delight, seals were everywhere. The gray

and white spotted blobs were scattered throughout the dune grass, popping their heads up with apparent annoyance as we passed directly beside them. The research staff referred to these young seals as "weaners," Sarah explained, partly because they've been weaned from their mothers but also because they closely resembled plump sausages.

During the gray seal breeding season, pregnant females haul out first, lumbering onto the beach in droves to seek out a decent spot to give birth. Each will bear a single pup. Gray seal pups are born with a thick coat of snow-white fur, or lanugo, that helps them to withstand the harsh winter temperatures, but they're bound to a terrestrial life until they pack on the pounds and grow their adult fur. Gray seal mothers nurse their pups for roughly three weeks while protecting them from aggressive males who have begun to come ashore eager to mate. During their lactation period, seal mothers will transfer roughly half of their body weight to their offspring through their high-fat milk. Then, when they're at the brink of starvation themselves, the mothers abandon their pups, unlikely to ever see them again.

The plump weaners then enter their own period of fasting. They survive off their blubber reserves until they shed the last of their lanugo, revealing a waterproof coat capable of insulating them from the cold North Atlantic waters. Eventually, instinct and hunger compel them into the ocean for the first time. There, they must learn to forage on their own, and quickly, as they'll be competing with tens of thousands of other seals for available food resources.

The seal mothers' struggles aren't yet over either, as the females must contend with any number of testosterone-fueled male suitors nearly twice their size. The males have also been fasting on the island, as they wait for any opportunity to mate. The females aren't able to put up as much of a fight as they might like, as they're quite weak at this stage, having sacrificed so much of their resources to ensure their pups' survival.

While gray seal pupping felt a bit extreme to me—mothers abandoning their offspring cold turkey after just a few weeks—it's fairly typical of other seal species. In fact, hooded seals, which breed on ice floes in the Arctic, win the award for the shortest nursing period of any mammal on earth. Pups are weaned off their mother's milk, comprised of 60 percent fat, between three to five days after they're born. Harbor seals, on the other hand, were more my speed—a gradual weaning process where mothers nurse their pups for four to six weeks while teaching them to swim and forage on their own. And when harbor seal pups get tired, they can simply hitch a ride on their mothers' backs.

After Sarah gave me a brief orientation to the island, we ventured out to take advantage of the limited time I had. I had arranged to meet with Nell den Heyer while I was on Sable, as she was on the island with her team carrying out their research. Sarah explained that Nell was finishing up a seal survey, but would join us later on our walk.

Sarah was in her midthirties and had a doctorate in feral horse population dynamics, which, naturally, is what drew her to Sable. The iconic equine residents were scattered across the island in small bands of four to five horses, usually with their heads down, muzzles half-buried in patches of dune grass, long, shaggy manes and tails aglow. They paid little attention to the seals, even those positioned directly underhoof.

Sable is only a mile wide at its thickest point. As we stood inland near the main station buildings, the dune cliffs on either side blocked a view of the ocean, but offered some welcome protection against the full force of the wind. Its strength was breathtaking.

Sarah led me along a narrow, well-worn trail through the dunes, which ran parallel to a wider sand road (of sorts) littered with seals. A mare nursing her foal grazed on the edge of a freshwater pond, one of several spring-fed ponds that helped to sustain life on the small island.

All around us, the dune grass swayed in rhythmic pulses. Waves of air swirled the dusty green blades in slow motion, like underwater seagrass beds. The connected force of the dunes seemed to be holding the island in place.

The sound of the wind overpowered any attempts at conversation along the trail. I felt like Charlie Brown listening to adults speaking in trombone. But the wind was merely a harmony. The melancholic melody was the collective vocalizations of hundreds of thousands of seals across the island—haunting, desperate, sorrowful coos. It was hard to imagine the sounds as the alluring voices of sea nymphs that captivated Odysseus and other lonely sailors. A massive flock of depressed pigeons might be a more appropriate comparison.

We headed toward the beach, moving slowly as I stopped every few seconds to ogle the seals. The largest of the males were sidled up beside females still nursing their pups, staking their claim, biding their time. Sarah explained that it wasn't the aggressive, testosterone-fueled males to worry about when it came to human safety on the island, it was the females protecting their young pups. Even a nine-hundred-pound bull is no match for a determined seal mother half his size. And a devoted seal mother wouldn't hesitate to lunge at a human she deemed a threat and sink her teeth into their shin to protect her pup. While the bite itself might not take you down, the resulting bacterial infection had a solid shot.

We passed by a particularly plump weaner, still in full lanugo, balanced on its side like an inflated balloon, rolls of fat gathered across its belly. "I try not to fat shame," said Sarah. "But let's just say that pup is going to be very productive in its life." A moment later, we approached a male seal attempting to mate with a feisty female. While he had managed to pin her under his massive frame, he seemed to have gotten turned around in the process of avoiding her sharp teeth as she fought back. Instead of mounting her in the standard mating position, he was lying on top of her with his belly to the sky. "I'm not quite sure what his game plan is there," said Sarah. "You're doing it wrong, man."

Sarah's radio buzzed at that moment and I could hear Nell's voice on the other end as Sarah explained our location. Seconds later, an ATV sped around a bend in the dunes, expertly dodging one unmoving gray seal after the next, each seemingly oblivious to the loud and potentially dangerous vehicle beside them.

Nell warmly welcomed me to the island before quickly debriefing with Sarah about the long-overdue arrival of groceries and equipment. They seemed particularly jazzed about a new supply of eggs. "That's just life here on Sable," Nell explained, turning to me. "But it is quite amazing that you made it. Sarah, you really worked some magic." "Dave isn't super happy with me for putting so much stuff on that helicopter," Sarah replied. "But look. You're all here now." I laughed, nervously.

Suddenly, a parade of five ATVs came zipping around the bend, dodging the scattered seals with the same agility as Nell. "Ah, here are the troublemakers now," she said. During the roughly six-week gray seal breeding season, the busiest time of year for Nell and her research team, there are typically at least ten staff working on the island at any given time—counting seals, applying various research tags and tracking devices, and engaging in a whole host of other fieldwork activities. Nell exchanged a few words with her team before they sped off in the direction of the main station. She then ditched her bike temporarily to join Sarah and me on our walk.

I peppered Nell with questions as we headed toward a gap in the dunes that offered a pathway to the ocean. But as we came upon another herd of seals, it became challenging for me to focus on anything outside of the chaotic scene unfolding before my eyes. The wind was even wilder as we neared the water, and the beaches were littered with seals. Just beside us, two enormous males were engaged in a bloody battle, their hulking frames rising vertically out of the sand. They growled and roared as they sank sharp, yellowed teeth into each other's already bloodied faces and necks. The males were careful to protect their hind flippers—sustaining injuries there would dramatically impact their ability to maneuver in the water, and ultimately survive. But it seemed faces and necks were fair game.

"They're just starting to come in now," said Nell, nodding toward the males. I realized I'd stopped talking mid-question and was now staring open-mouthed at the battle. "They're beginning to breed. So we drive the island systematically looking at the backside of every single seal." The team was looking for seals that had been branded or tagged to feed into their assessments of the behavior and status of the population. As Nell explained, the population of gray seals on Sable Island had been growing at its maximum rate up until the mid-1990s, which meant about a 14 percent population increase each year. After the mid-nineties, the rate of growth began to slow, but it wasn't until their 2021 survey that the population appeared to have leveled off. "It was the first time we'd done a survey in sixty years where we didn't have more seals on the island, more pups being produced."

I asked Nell what might have caused the population on Sable to stop growing. Competition was the primary hypothesis, she explained. Some of the tracking work her colleagues have done on young seals has suggested they forage in the same areas as the adult females, but they get pushed to the side and have to take longer trips. "We do know that it's the young animals that are having trouble," she said. "Juvenile survival has really gone down." At the same time, you can't ever really know what's happening, she added. The data are limited, and the researchers must make inferences about the population based on the information they gather.

While Sable's gray seal numbers appear to have leveled off, some of Canada's coastal gray seal colonies in the Gulf of St. Lawrence, as well as in Southwest Nova Scotia, are continuing to grow, along with the New England gray seal colonies. "The New England story is really interesting," said Nell. Gray seals moved out to truly remote areas like Sable Island after being hunted out of their historic territory—not just in the U.S., but in Canada as well—and now they're recovering across their entire range.

One of DFO's research aims, and the reason Nell and her team are deployed to Sable Island each year to collect population data on seals, is to inform Canada's commercial gray seal hunt. While Americans are banned from hunting seals under the terms of the Marine Mammal Protection Act, Canadians continue to operate a commercial seal hunt. Seals aren't harvested on Sable Island, but a government-regulated gray seal hunt occurs each year in the Gulf of St. Lawrence. And in 2023, the government expanded the harvest region to the Scotian shelf, Nell told me. "So we provide harvest advice, but the harvest is very small," she explained.

But there's another, perhaps even more significant purpose to their research. "The biggest interest in gray seals is their impact on fish stocks," said Nell, "which is why we do a lot of diet work." The fishing industry in Atlantic Canada is vital to both the economy and culture. Questions around what seals are feeding on and how they're influencing fish populations, particularly cod, has prompted the Canadian government, along with many universities and other research groups, to devote significant funds to investigate.

By this time, we'd circled around a section of the beach and began walking back inland toward the opposite side of the island.

As Nell needed to get back to her afternoon fieldwork, she said goodbye and headed off to find her bike, while Sarah and I continued toward the beach.

The entrance to the water was sandwiched between two towering dune cliffs. But as we got closer, we discovered an army of seal mothers, pups close to their sides, lined up in tight formation blocking the path. The females were flanked by dozens of testosterone-fueled males.

Assuming we had no choice but to turn back, I was already mid-pivot when Sarah calmly said, "We should probably find you a walking stick." Before I could ask how a walking stick could possibly protect us against an army of aggressive seals, Sarah leaned forward and snatched one from the seemingly barren sands, then handed it back to me. *I suppose there are worse ways to die*, I thought, white-knuckling what I assumed was a poking device defense system.

We moved slowly through the seal gauntlet, greeted by a chorus of angry hisses, growls, and snarls of disapproval. I held my breath as we approached the narrowest part of the path, where we'd be forced to walk directly between multiple sets of mothers and pups positioned mere feet from one another. "Best to stay close here," said Sarah. She turned briefly to look back, and was likely startled to realize I was already standing inches from her. At that moment, one of the seal mothers, teeth bared, lunged forward toward Sarah. I held my breath—this was it, cause of death: seal riot, unsuccessful breach of seal defense line.

But instead of retreating at top speed (my preferred course of action), without a word, Sarah raised her walking stick over her head and slammed it into the sand between her and the furious female. Like Moses parting the Red Sea, the move was miraculously effective. The seal halted her advance mid-lunge, while continuing to hiss and growl as we passed. With Sarah's staff defending our left flank, I realized that I was now in charge of our right, just as another seal mother geared up for an attack. I attempted to mimic Sarah's move, but my "staff slam" was more of a gentle sand poke, followed by a confused dragging of my stick, which I then nearly tripped over. If nothing else, my actions seemed to convince the seal mother I was too clumsy to be a threat, and she too halted her advance.

Our final hurdle was to pass through a dozen males that had gathered on the outer edge of a tide pool, but thankfully, there was little time for panic. As soon as we approached, the males immediately turned and high-bellied it into the center of the tide pool, well out of our path. *Typical male aggression*, I thought, *nothing but vocalizations.*

The wide beach was filled with seals, blubbery blobs scattered across the sand as far as you could see—their voices amplified by the deafening winds.

With about an hour left until my scheduled departure, Sarah led me to the top of one of the tall dune cliffs overlooking the water toward the east end of the island. On reaching the top, breathless, my hiking boots filled with sand, I looked down at the seals hauled out on the beach below. "Be careful with your footing," warned Sarah as I inched closer to the edge. "The dunes erode—it would be a deadly drop."

A band of wild horses calmly grazed nearby, the wind swirling their long manes as if they were being filmed for a L'Oréal commercial. Behind them, waves rose and crashed along the shoals, seeming to extend for miles. The eastern tip of the island curled out into the sea, the lines between land, sea, and sky blurred into one. I was so absorbed in the otherworldly scene that I jumped at the sound of Sarah's voice. "I tell you what," she said. "I feel like you can't have the Sable Island experience unless you get to experience it on your own. You have a watch on you?"

I looked at her, confused. "We really need the helicopter lifting off at three thirty p.m.," she said, "so just make sure you're back by then." She nodded toward the main station, visible in the distance. "Enjoy," she said, smiling, as she turned to descend the sandy slope.

I looked over the edge of the cliff, overwhelmed by the sheer number of seals scattered across the sand. As I watched a group of males resting in the surf, I noticed a female, a short distance down the beach from the males, rapidly galumphing toward the water. Unfortunately, I wasn't the only one to notice. Periscoping their massive heads toward the sudden commotion—the female was clearly trying to make a break for it—the males immediately began lunging over one another in a desperate, instinct-driven struggle to gain the advantage. The water churned and splashed as the males rushed toward her, one swimming beneath a wave, others porpoising through the surf, each fighting to be the first to arrive. I flashed back to a gray seal documentary I'd watched months earlier and recalled a graphic scene where a mob of males attempted to mate with a single female and ultimately crushed her to death. I turned my back to the scene, focusing instead on a band of peacefully grazing horses. I realized I didn't have the stomach to find out if she made it.

But I was frustrated with myself for that very notion. I knew rationally this was simply nature, raw and unedited. *The plight of an individual animal is a necessary sacrifice to secure the health of the species*, I thought, while channeling the reassuring voice of David Attenborough. These are merely wild seal instincts at play. But I couldn't help but think that it's a cruel fate to be a female gray seal on Sable Island.

AS MUCH AS MY research had prepared me for the reality of the gray seal breeding season on Sable, part of me had still expected, or rather hoped for, more of a happy seal beach party. Cuddly, white-coated pups belly hopping around the island, playing in the sand beside their loving, devoted mothers. And sure, I knew mating would be a bit graphic at times, but I'd pictured more of an "opt-in" seal sex party (despite Greg's warnings at the hangar that morning). Instead, I'd witnessed National Geographic on steroids. The seals didn't come to Sable to socialize, they came to further their species. Pup, nurse, mate, survive. The plump, wide-eyed pups were fearful. Their mothers were deflated balloons, weakened by their sacrifice. The males were battle-scarred and desperate.

As I'd walked with Sarah that afternoon, there were more seals than I could possibly count, but not all of them were alive. The dead seals I spotted were mostly pups and females, some of which had been heavily scavenged by gulls. The line between life and death for a gray seal on Sable Island is razor-thin, and at times seemed to disappear altogether, swallowed up by the rapidly shifting sands.

Before Nell returned to her research team, we stopped to observe a female seal resting beside her pup with a rake-like wound on her back. While the skin had mostly healed, the marks were still quite prominent. "Probably a shark bite," she explained. I asked her if she could tell which species from the marks, and whether this might be related to the corkscrew seal wounds. Nell smiled. "Ah, yes. Everyone's always curious about the Sable Island sharks."

IN 2010, WHEN ZOE Lucas and Lisa Natanson published their paper proposing Greenland sharks as the culprits behind the corkscrew seal deaths, scientists on the other side of the North Atlantic had arrived at a very different theory. It turned out that Sable Island wasn't the only place mutilated seals had been washing up—seals with nearly identical wounds had been found along beaches in the U.K. David Thompson, a scientist at the Sea Mammal Research

Unit at the University of St. Andrew's, hypothesized that the lacerations were being caused by a ducted propeller, which was used to slow ships like tugboats and barges. To test the idea, the researchers created small wax models of seals that they fed into a scale model of the ducted propellor. Sure enough, the marks on the wax seals were nearly identical to the corkscrew wounds.

In response to the findings, in 2012, the U.K. government issued a formal advisory for any ships with ducted propellers to avoid seal conservation areas during breeding season.

But two years later, a shocking research discovery threw both the Greenland shark and ducted propeller theories into question. In December of 2014, Amanda Bishop, a doctoral candidate at Durham University, was observing a gray seal colony on Scotland's Isle of May when she witnessed something she'd never seen before. A large male seal dragged a young pup into a shallow pool, drowned it, and then used his teeth to tear into the pup's flesh and feed on pieces of its blubber. The flesh ripped along a clean collagen line at a roughly 45-degree angle, resulting in a distinct corkscrew wound identical to those that had been observed on both sides of the Atlantic. The same male seal was observed conducting this act four more times that same week. Nine other seal carcasses were discovered in the same area exhibiting similar wounds, likely caused by this same male.

The behavior, known as "cannibalistic infanticide," is common in some species, including chimpanzees, as a means of reducing genetic competition, but it had never before been observed in gray seals. In response to these new data, the U.K. government promptly rescinded their advisory to the shipping industry. Some conservationists were frustrated by the government's decision, as they considered ships to be a continued threat to seal populations, but the data, which included video footage and a series of photographs, were hard to dispute.

Meanwhile, scientists searched for answers as to *why* gray seals might exhibit this behavior. Some suggested it may have been an effort to gain energy while on the island, as males typically fast during the breeding season. Feeding on a pup's rich blubber would enable an adult male to stay longer in the colony and potentially mate with more females. Others disagreed, as it didn't explain why the seal killed so many pups, and yet consumed so little of each carcass. Perhaps there was no ultimate purpose, it was just the work of a few rogue seals with something not quite right in their heads, sort of like human serial killers.

Despite the U.K. research, scientists on Sable Island have yet to observe adult male gray seals attacking gray seal pups or harbor seals. Without direct evidence, the jury was still out on the Sable Island seal killers, and it seems that Greenland sharks haven't yet been ruled out as a suspect. Zoe herself emphasized the uncertainties. "It's a theory," she said. "It absolutely is a theory, as it hasn't been actually observed. So basically, it's by a process of elimination and circumstantial evidence pointing to Greenland sharks." She added that it was possible that a small proportion of the wounds may have been caused by cannibalizing adult gray seals.

I asked Nell for her opinion on the prevailing theories. She seemed to gravitate toward the male gray seal hypothesis, given the video evidence from across the pond, but without proof on Sable Island, she said, "It's hard to know." Aside from the lingering questions around the culprits and motives, it remains unclear why the corkscrew seal cases have only recently emerged.

The most fascinating aspect of the corkscrew seal saga, to me, was what it revealed about the limitations of science, and the mysteries of creatures that spend so much of their lives underwater. Even on this small, isolated island where researchers study the seals year-round, there seemed to be a limitless supply of new discoveries. And while the corkscrew seal research was certainly gruesome, it was also a poignant example of the complex nature and behaviors of seals themselves. Before my visit to Sable Island, I'd begun to think of seals less as wild creatures, and more as charismatic sea dogs—some talk, others perform tricks or fling snow with their flippers. They were sassy and playful, but generally docile. It was the story of seals we learned as children, while holding fast to our smiling seal stuffies from the aquarium gift shop.

But nature is merciless. Sable Island offered a window into natural animal instincts and processes—predation, competition, survival, the innate and desperate struggle to simply exist in this world. It was raw, ruthless, wild. It was beautiful and complex. I searched for meaning in the chaos.

Toward the end of my time on Sable, I came upon two weaners huddled beside each other on a high dune cliff. On hearing my approach, the smaller of the two turned to stare at me with wide eyes that radiated fear—or at least, that's what I perceived.

Anthropomorphism, the practice of attributing human traits and emotions to animals, is considered a sin of sorts among many scientists, as well as science journalists. But the primatologist Frans de Waal coined a new term: anthropodenial. In a 1997 essay in *Discover Magazine*, de Waal defined anthropodenial as "a blindness to the humanlike characteristics of other animals, or the animal-like characteristics of ourselves." In our effort to build a barrier between ourselves and "other" animals, we risk overlooking the intentions or emotions of nonhuman species. "In the end we must ask: What kind of risk are we willing to take—" wrote de Waal, "the risk of underestimating animal mental life or the risk of overestimating it?"

I spent my fortieth birthday the next day in my Halifax hotel room nursing one of the worst hangovers I'd had in years, which ironically wasn't caused by alcohol. In the excitement of my Sable Island adventure, I had neglected to consume a single drop of water or bite of food during the entire journey. But it was worth it. I couldn't imagine a better way to ring in a new decade than with 400,000 seals.

8

Scamps or Scapegoats?

I WAS SEVENTEEN WHEN I kissed my first codfish. It was evening and I was knee-deep in the harbor of a remote fishing village on the northeast peninsula of Newfoundland. The icebergs floating on the horizon had long faded into the darkness. When the "Master of Ceremonies" lifted the fish to my face, I kissed what I assumed were its lips, in fish terms. It was clear by the stench that this was no fresh catch, so my standing in the frigid water was symbolic at best, or at least provided an added layer of delight for my Newfie friends, who made little effort to stifle their laughter. It was the final step of my "screeching-in" ceremony, a tradition where foreigners perform a series of rituals, much to the amusement of the locals, to become an honorary member of "The Rock."

It was the summer before my senior year in high school, and I had journeyed to Conche, Newfoundland, for an internship with my dad's environmental nonprofit, the Quebec-Labrador Foundation. My mission was to support local students and residents working with the French Shore Historical Society as they built the groundwork for a new ecotourism industry. For a naive teen from Massachusetts, Conche was a world beyond my imagination. Cliff trails meandered past tumbling waterfalls, humpback and minke whales spouted plumes of sea spray off the coast, a harp seal perched on an iceberg just beyond the harbor, hitching a very slow ride to warmer waters. For at least a week in July, a beluga mistaking a white-hulled dinghy for its mother was the talk of the town.

Newfoundland's northeast coast was, and is, a gold mine for nature-loving tourists, but the remote nature of rural villages like Conche could be a challenge, even for intrepid travelers. But perhaps the bigger challenge was that not all residents were eager to embrace a new tourism industry. These were Newfoundland fishing towns, through and through, except for one issue: there weren't any fish.

Newfoundland is called "The Rock" for a reason. Between the lack of fertile soil and the short growing season, agriculture is nearly impossible. But the island was, at one point, surrounded by one of the most fertile fishing grounds in the world. And the king of fish—the species that drew the Vikings, the Basques, and later, European settlers—was Atlantic cod. For centuries, cod were so abundant—not only on the Grand Banks of Newfoundland but further south in the Gulf of Maine and Georges Bank off New England—that it was said you could walk across the water on their backs. According to Mark Kurlansky, author of *Cod*, the North American coast ". . . was churning with codfish of a size never before seen and in schools of unprecedented density, at least in recorded European history."

Cod fishing was a way of life. The screeching-in ceremony I was cajoled into gave honorary Newfoundlanders the chance to pay homage to the species around which Newfoundland and Labrador built its economy and its vibrant culture.

For centuries, fishermen built their livelihoods around the seemingly limitless Atlantic cod. But in the 1950s, the arrival of industrial fishing fleets transitioned an already heavily fished ecosystem into a wasteland. In a matter of decades, factory trawlers with fish finders and freezers hauled up entire schools of cod, as well as anything else that got in the way. They demolished the stocks with alarming efficiency.

By the early 1980s, Newfoundland's inshore fishermen, experiencing dramatic declines in their catch, sounded the alarm that the cod were disappearing. But the government failed to respond. The modeling projections developed by government scientists showed the cod were doing fine. It was later determined that these models had been based on flawed assumptions, but by then it was too late. The population of cod on the Grand Banks, the stocks that supported the largest fishery in the world, had been reduced by 99 percent.

In 1992, the Canadian government had no choice but to shut down the cod fishery. While barricaded in a conference room in St. John's, Newfoundland, John Crosbie, the fisheries minister, made a televised announcement about the closure in front of a panel of reporters, while hundreds of furious, and frightened, fishermen banged on the doors, demanding answers. How could it have come to this?

Overnight, more than 30,000 people lost their jobs. The fishing moratorium was expected to last two years, with the fishery reopening in the spring of 1994. But when I arrived in Conche in the summer of 2000, the cod had shown no signs of recovery, and the fishery remained closed.

Conche was one of many coastal communities in the region devastated by the fishery's collapse. By 1995, the unemployment rate in Newfoundland and Labrador was over 20 percent. In the first ten years after the moratorium, Newfoundland lost 10 percent of its population to outward migration. The Canadian government began a resettlement campaign, offering residents in some of the most remote fishing villages in the province a payment to voluntarily relocate to towns the government had identified as "growth areas."

In response to the loss of the fishery, some fishermen invested in new gear, rejigging their boats to target other species. Without cod as predators, the populations of snow crab and northern shrimp were thriving. Others were able to find work in other industries. The father of my host family in Conche, who had fished for decades, was forced to sell his fishing license and take a job at a local sawmill. He embraced the potential of ecotourism, as did my host mother, both eager to share Newfoundland's vibrant culture and heritage with the world. But every so often, he'd play his accordion for us and sing sea shanties about a way of life that no longer existed. I was a seventeen-year-old summer intern, and only beginning to understand the rich culture, history, and natural beauty of Newfoundland. But those wistful moments left an impression. The sense of fury and heartbreak over what had been lost, and the hope for a future that had yet to be written, were striking.

By the mid-1990s, with the ground fishery still closed, the Canadian government made a highly controversial decision, infuriating animal welfare activists and environmental groups around the world. At the same time the cod had disappeared, the population of seals in Atlantic Canada, which included harp, hooded, gray, and harbor seals, had dramatically increased. Some blamed seals for consuming too much cod and preventing its recovery. To control the growing populations of seals, the Canadian government expanded the commercial seal hunt by offering subsidies to hunters.

THE COMMERCIAL SEAL HUNT off the northeast coast of Newfoundland and in the Gulf of St. Lawrence dates back to the

1700s. The primary species of interest has historically been harp seals, and to a lesser extent hooded seals, both of which gather annually on pack ice nurseries. But gray and harbor seals were also hunted during this time. By the 1840s, nearly half a million harp seals were being killed each year for their oil and highly valuable furs. Of particular interest to the thriving European fur market was the soft, white fur belonging to harp seal pups less than two weeks of age.

By the early 1970s, intense hunting pressure had reduced the population of harp seals to a small fraction of its historic size. Around that same time, animal welfare groups, led by Greenpeace and the International Fund for Animal Welfare, launched a media campaign to spotlight some of the inhumane hunting practices they'd witnessed on the ice. Activists began arriving at the hunting grounds armed with video cameras to document the event. What was once a clandestine affair, occurring on ice floes far off the Canadian coast, was suddenly thrust into public view. International news outlets began publishing images of days-old seal pups being clubbed to death on blood-soaked ice.

The American public was outraged by these images. Lee Talbot, a renowned ecologist and geographer who authored the draft legislation that became the Marine Mammal Protection Act, cited the Canadian seal hunt as one of the key "trigger events" that enabled the U.S. government to pass such monumental marine mammal protections. The photos of "baby seals" being bludgeoned to death on Canadian ice floes tore into the hearts of Americans. "Friends in Congress told me they had never received such a large volume of letters on any subject, other than the Vietnam War," wrote Talbot in a 2017 opinion piece for *The Hill*. It's worth noting that amidst our collective moral outrage, there was little focus on our own role in killing seals through decades of bounty hunts.

Despite the public relations nightmare, the Canadian government refused to shut down the seal hunt, so the activists shifted their attention to Europe, pressuring the European Economic Community to ban the import of seal products. To increase their reach, they enlisted the support of celebrity activists. In 1977, French model Brigitte Bardot posed on the ice while holding a furry white seal pup, an image that immediately captured the media's attention. (In later years, Paul McCartney and his then-wife, Heather Mills, posed for a similar photo.)

In 1983, the European Community banned the import of whitecoat seal fur. Seal products from animals just over two weeks of age weren't included in the ban, but increased public awareness about the nature of the hunts had shaken the European fur markets. Demand for seal products effectively disappeared.

While the anti-sealing campaigns were focused on the commercial hunt occurring off the coast of Newfoundland and in the Gulf of St. Lawrence, in the minds of foreigners, all Canadian seal hunts were the same. Animal welfare groups had made little effort to distinguish between seal hunting practices, species, and regions—the implications of which proved to be devastating.

In the Canadian Arctic, and in Arctic areas around the world, Indigenous people have long depended on seals for their survival. Seal meat has been an essential food resource for Inuit communities, and sealskins are made into boots, pants, mittens, coats, and other clothing for warmth and to protect against wet conditions. But the purpose of the Inuit hunt, the methods used to kill the animals, and even the species hunted differed dramatically from the Canadian commercial hunt. Inuit hunters travel across the ice in search of seal breathing holes made by ringed seals. The hunters wait, in freezing temperatures, for a seal to return to its hole, then swiftly kill the animal and transport its carcass back across the ice.

Peter Irniq, the former Commissioner of Nunavut, a vast territory in northern Canada stretching across most of the Canadian Arctic, described his experience growing up seal hunting with his family in the book *Sacred Hunt,* authored by David Pelly. Irniq writes:

> *When my father returned home with a seal, it was my mother's responsibility to butcher the animal. As soon as the seal was pulled through the entrance of our iglu, my mother would take her ulu (a moon-shaped, woman's knife), and get ready to skin the seal. Before she did that, she would take a small piece of freshwater ice and gently put it into the seal's mouth and let the dead seal drink water. She would then say, translated into English, "This is so that all seals under the ice will not go thirsty." How powerful and strong this simple message to the spirits was. And spirits listened!*

While the seal hunts were critical for subsistence, Inuit communities relied on commercial markets to support their other essential needs. The money earned from the sale of seal products was a core component of their economic welfare, and the loss of the European market upended their lives and livelihoods. While the bans had included exemptions for the Inuit, the reputation of seal products from Canada had been tainted, and demand had dried up. Some activists publicly voiced their support of subsistence hunting, but that did little to address the sudden and overwhelming loss of income.

In 1978, Paul Watson, a prominent animal rights activist who had recently been ousted from the Greenpeace Board of Directors due to conflicting views, spoke candidly during a CBC Radio interview about how images of harp seal pups had been exploited to support unrelated species and causes. Harp seals have glands that keep their eyes lubricated to protect against salt water, and their lack of tear ducts can make it appear the seals are crying. As Watson attested, these images of crying "baby seals" were essential to Greenpeace's broader fundraising efforts for endangered species, despite the fact that harp seals were not endangered. "I think that of all the animals in the world, or any environmental problem in the world, the harp seal is the easiest issue to raise funds on," said Watson. "It's an image that goes right to the heart of animal lovers all over North America." Yet though he appeared to express remorse during this radio interview, in her 2016 documentary, "Angry Inuk," Inuit filmmaker Alethea Arnaquq-Baril pointed out that Watson employed the exact same tactics after founding the Sea Shepherd Conservation Society.

After the European market collapse in the 1980s, the Canadian commercial seal hunt continued, but the number of seals killed each year was significantly lower. There simply wasn't enough money to be made. But the government continued to periodically cull seals to, as they attested, support the fishing industry by reducing their competition.

But in 1995, three years after the closure of the cod fishery, the Canadian government announced they would offer financial subsidies to sealers as a way to renew interest in the hunt. The year before the subsidies were enacted, fewer than 66,000 harp seals were harvested off Northeast Newfoundland and in the Gulf of St. Lawrence. A year later, that figure nearly quadrupled, to more than 240,000 seals. The subsidies worked.

While the government pointed to the harmful impacts of seals on cod stocks, there was no scientific data to support this assertion, outside of the simple fact that seals eat cod. How much cod were seals consuming, where exactly were they consuming it, and what impact did that consumption have on the cod population as a whole? There were no answers to these questions, nor were there any data to suggest that killing seals would hasten the return of cod. It certainly hadn't yet.

Since then, the commercial seal hunt has continued. But in 2012, four independent marine scientists—Hal Whitehead, Sara Iverson, Boris Worm, and Heike Lotze—published an open letter in response to a gray seal cull proposal, stating that "The scientific community is in agreement that the principal cause of fish stock depletions is human fisheries and that there is no credible scientific evidence to suggest a cull of grey seals in Atlantic Canada would help depleted fish stocks recover. Similar measures taken in the past have never succeeded to increase fish stocks."

The scientists explained that the primary reason cod have failed to recover is because of their natural fish predators, not seals. Culling gray seals, they argued, would have a negative impact on cod, as seals prey on the fish that consume cod eggs and young cod. "An uncontrolled cull of grey seals is not an experiment, would teach us almost nothing about the interactions between seals and fish and would be a waste of valuable resources and animal lives," they wrote.

THE CRISIS WITH THE cod fishery was also playing out in important ways in the U.S. While the cod stocks in Georges Bank and the Gulf of Maine were much smaller than those that had existed on the Grand Banks of Newfoundland, cod had historically been a valuable resource and source of identity in New England, even giving Cape Cod its name. A painted "sacred cod" has been hanging in the Massachusetts State House since 1784 as a symbol of prosperity.

But in the 1980s, as cod stocks in Newfoundland were showing signs of massive decline, the same was happening in New England. Jill Bubier is a retired professor of environmental science at Mount Holyoke College. In the mid-1980s, she was working for the Marine Law Institute at the University of Southern Maine. As an environmental attorney at the time, part of her role was to research and report on fisheries management decisions, which meant attending meetings held by the New England Fishery Management Council. The council was a regulatory group made up of government scientists and fishermen determining policies for the fishery. Jill's assignment was to observe the meetings and help interpret the legal implications of the decisions.

"But the whole time I was at those meetings, I kept thinking, these people have really different agendas," said Jill when we spoke on the phone. The fishermen had a shorter-term perspective, she explained. They were focused on putting food on the table for their families over the next five years. The biologists, meanwhile, were focused on the longer-term—how to manage resources to ensure they'll still be around in twenty or thirty years. It would come down to arguments between the two groups over fishing restrictions. "In the end, there would be some compromise, but it didn't really work." In Jill's view, the management structure just wasn't designed to preserve the fishery.

While no one knew for certain just how dire the situation was at the time, looking back today, Jill said the writing was on the wall. The biologists would present their stock assessments at the meetings, emphasizing how significant the population declines were. "They were sounding the alarm," said Jill. But the fishermen often shrugged off the data, explaining that they were still catching a lot of fish. In the end, the scientists and fishermen would negotiate on gear restrictions or seasonal fishing restrictions, or whether to issue fewer fishing licenses that year. "But it was just nickel-and-diming," said Jill. "It wasn't dealing with the real problem."

SCIENTISTS IN THE U.S. and Canada have been studying seals' diet and foraging behavior for decades using a variety of research methods, including studying the stomach contents of dead seals, analyzing their scat, and observing their feeding habits in the wild. Marjorie Lyssikatos is a fishery biologist at the Northeast Fisheries Science Center in Woods Hole, Massachusetts. Part of her work, along with her now-retired colleague, Fred Wenzel, has involved sampling the stomachs of gray seals mistakenly caught and killed as bycatch in New England ground fisheries, to better understand their diet.

But, as Marjorie explained to me on a call, seal diet research is tricky, and the results can vary depending on factors like the age of the animal, where it was foraging, or even the research method used. Broadly speaking, seals are highly opportunistic. "Both gray and harbor seals appear to be generalists," said Lisa Sette, a marine biologist with the Center for Coastal Studies, who I spoke to on a call. "They appear to take advantage of what is seasonally abundant during the time of year that they forage." According to some of Majorie's preliminary findings, there's consistent evidence of red hake, silver hake, and sand lance in the diets of gray seals. For harbor seals, the predominant species are redfish, shortfin and longfin squid, and herring.

When it comes to cod, said Marjorie, "it's an important prey item when it's available. There's no reason to believe if cod were very abundant in the environment that seals wouldn't try to capitalize on it. They have in the past and I don't see why they wouldn't continue to do so."

But the cod question gets even trickier the deeper you dive. Some suggest that seals may actually be *benefitting* cod's recovery.

Seals prey on forage fish species like hake and sand lance, which consume cod eggs and juvenile cod, meaning that seals may be helping to control the predators of cod. "It's very reasonable to believe that even with the recovery of these seal species, the predominant prey in their diet are actually relieving pressure on cod," said Marjorie.

With all of this noise in the data, there doesn't seem to be any clear evidence that "thinning the herd" will have a positive impact on fisheries, and most of the research has been inconclusive. Not only that, the logistics of a seal cull are wildly complex. In 2009, the Canadian government commissioned a study to evaluate the feasibility of a gray seal cull on Sable Island. Linda Pannozzo, a journalist based in Nova Scotia and author of *The Devil & The Deep Blue Sea*, obtained the results of the analysis through an Access to Information request.

According to Pannozzo, when it came to killing the seals, the logistics "would be gruesome, at best . . ." The culls would need to occur during the winter breeding season, reported the firm. The adult seals would have to be killed with rifles, and the pups with either a rifle or by clubbing. A considerable amount of equipment would be needed, including roughly twenty mobile crematoriums to quickly incinerate the carcasses. Otherwise, the firm noted, ". . . the onset of rot and disease would be fast, resulting in biological hazards and health and safety issues for the workers." If the cremation couldn't be carried out on the island, the carcasses would need to be transported to the mainland, involving helicopters slinging dead seals from shore onto a supply ship, plus truck convoys on the mainland to transport the seals. Sterilization was also considered, but it was a less favorable option, reported Pannozzo, as it would have a far less immediate impact on cod recovery.

Meanwhile, new data have emerged suggesting that climate

change may be playing an even more significant role than previously thought when it comes to cod. In 2015, Andrew Pershing, then the chief scientific officer at the Gulf of Maine Research Institute, and his colleagues published a study in *Science* linking the collapse of cod to rapidly warming ocean waters. The challenge was that even with strict quotas set for the New England cod fishery, fishery policies aren't accounting for the role of climate change in increasing cod mortality and reducing reproduction.

Scientific data (or lack of data) aside, the political, economic, and social pressures around the need to "manage" seals are significant. While the Canadian government is taking action to reduce seal populations with the goal of supporting fish recovery, the U.S. government has no such authority under the terms of the Marine Mammal Protection Act.

But that doesn't mean U.S. fishermen are any less frustrated than Canadian fishermen. Seals have a preference for fatty cod livers, so on top of consuming a lot of fish, they've gained a reputation for being "wasteful eaters" (not unlike many of us humans).

AS THE MONTHS OF snow and ice reluctantly yielded to the wet, cold, gray days they call spring here in Maine, I'd already set my sights on a new seal research destination: Washington State.

It seemed there were parallel stories playing out on the East and West Coasts. In New England, seal-fishery conflicts were centered on the recovery of gray seals and Atlantic harbor seals. In the Pacific Northwest, the conflicts were centered on California sea lions and Pacific harbor seals. But the general story—the depletion, protection, and recovery of pinnipeds, followed by intense conflicts with humans—was remarkably similar. I was hoping there was something relevant I could learn from the West Coast experience to better understand the situation in my own backyard.

What I didn't know at the time was that I was about to discover one of the wildest stories of pinniped shenanigans I'd ever heard. It was the story of a group of clever, hungry, and ill-fated sea lions, a team of government biologists, known to some as the "blubber busters," and a costly effort to save two of the most valuable, and threatened, fish in the U.S.—Pacific salmon and steelhead trout.

PART THREE
SPRING

> People here used to believe
> that drowned souls lived in the seals.
> At spring tides they might take shape.
>
> —SEAMUS HEANEY, "THE SINGER'S HOUSE"

MAP OF PUGET SOUND, WASHINGTON

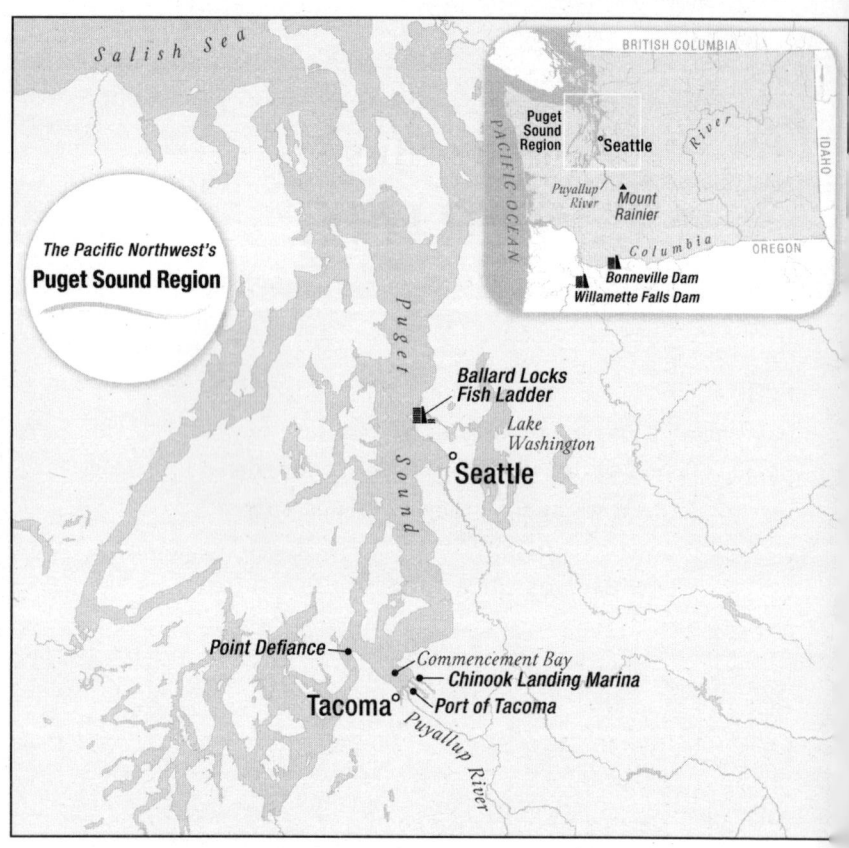

9

Blubber Busters

THE GRAND OPENING CELEBRATION of Seattle's Ballard Locks, the man-made gateway connecting the eastern lakes of Washington to the Pacific Ocean, was held on July 4, 1917. Crowds gathered, bands played, and one of Boeing's first aircrafts flew overhead as a parade of boats passed through the newly constructed aquatic elevator in the heart of the city. Former President Teddy Roosevelt sent a telegram congratulating the city of Seattle for the achievement.

The locks consisted of a holding pen of sorts positioned on one end of a dam. As boats and barges lined up in the canal, they took turns entering the locks. Once the boats were in position, bells rang and the entrance gates closed behind them, locking them in. Operators from the Army Corps of Engineers then opened a series of valves, releasing water as if emptying a bathtub. The boats descended roughly twenty vertical feet, at which point the exit gates opened and the boats motored out as they continued on into Puget Sound.

The locks were a boon for business, transforming Seattle into a major U.S. port city. The ambitious engineer who led the construction of the locks, Hiram M. Chittenden, had solved a major challenge plaguing the state of Washington: how to transport its seemingly limitless natural resources, including coal, timber, and gravel, from the Eastern part of the state to the Puget Sound, where it could be exported.

What officials didn't realize at the time, however, was that the Ballard Locks would eventually become a battleground for a feud between humans and sea lions (and more recently, humans and seals) over one of the most important resources in the state of Washington: fish.

Constructing the locks had required drying up wetlands and rerouting multiple rivers to create enough water flow to operate the new system—rivers that supported significant runs of salmon and steelhead trout. These fish are born in freshwater rivers and streams, then migrate out into the ocean, where they live for years before returning to spawn in the very same waters in which they were born. But the construction of the locks altered these free-flowing river habitats, blocking the fish migrations.

To address this issue, Chittenden included the design of a fish ladder in his plans for the locks to allow passageway for the fish. The ladder was comprised of a series of steps on a gradual incline. It enabled the fish to swim from one step to the next as they made their way upstream to their natal waters to spawn. The fishway was effective, and hundreds of thousands of salmon and steelhead passed through the locks each year.

But eventually, a problem emerged.

In the early 1980s, a few opportunistic pinnipeds began appearing at the locks during the seasonal gatherings of fish. In their natural habitat, salmon and steelhead can find places to hide from predators along the banks of the river. But the flat, concrete walls of the canal leading to the locks offered no such protections. The situation was even worse when the fish reached the ladder. The ladder forced massive numbers of fish to funnel into a narrow opening at its base, wriggling furiously as they squeezed into the small passageway.

For the pinnipeds, the congregation of fish in a single, easy-to-access location was an opportunity too good to pass up. Over time, a few of these fish-loving marine mammals began positioning themselves at the base of the ladder during the runs, picking off and swallowing one fish after the next. But what began as an entertaining performance for the visiting public soon became a major concern for fisheries managers, fishermen, and Native American tribes who relied on these fish to support their economies and cultures.

Prior to the 1900s, an estimated 7.5 to 16 million adult salmon and steelhead returned each year to Pacific Northwest tributaries. Today, the numbers of wild fish returning have declined by well over 90 percent by some estimates. The fish have been devastated by a host of human-caused problems throughout the past century—development, pollution, warming waters, overfishing, and barriers, including dams, that impede their migrations.

Protecting these fish wasn't just important to residents of Washington State—it was a federal priority. Of the roughly $1.2 billion spent on all threatened and endangered species in the U.S. in 2020, over *half* of these funds went to support the recovery of Pacific salmon and steelhead.

As the fish populations have declined, the impact of predators such as Pacific harbor seals and California sea lions, whose numbers have been on the rise in recent decades—a recovery story that mirrors that of the East Coast seals—appears to be taking a bigger toll on the already depleted fish stocks.

The construction of dams across the Pacific Northwest, which funnel fish into "pinch points" like the ladder at the Ballard Locks, presented a unique challenge for wildlife managers who were tasked with protecting marine mammals *and* managing fish stocks. At the Ballard Locks, pinnipeds had discovered an all-you-can-eat seafood buffet. How could officials prevent a few opportunistic, and federally protected, marine mammals from wiping out entire populations of fish?

What happened next inspired one of the most controversial amendments to marine mammal protections in the U.S., and it all played out in the heart of downtown Seattle.

IN MID-APRIL, SHORTLY AFTER arriving in Seattle, I headed south down Interstate 5 to Point Defiance Park in Tacoma, where I'd arranged to meet with a man by the name of Steve Jeffries. Now retired, Steve was a former marine mammal research scientist at the Washington Department of Fish and Wildlife. In addition to his marine mammal knowledge and experience, I was hoping to learn more about his role in what was arguably the state's biggest pinniped conflict of all time.

I wasn't sure how hard it would be to find Steve among the park's many visitors that morning, but I shouldn't have worried. There was only one Jeep in the lot with the license plate "BLUBBER." Steve was in his early seventies, tall with gray hair and a beard to match. As we walked along a pedestrian path by the water, he recounted some of his early work with harbor seals. We approached a breakwater where harbor seals occasionally hauled out at low tide, although there were none in sight that morning.

"Look out over Puget Sound," said Steve, nodding toward the water. I followed his gaze. To our left, deep blue water surrounded emerald forested islands and peninsulas, with large waterfront homes nestled between the evergreens. To our right was the Port of Tacoma. Industrial-sized vessels crowded in the harbor while smoke plumes billowed from factories. Behind the thick veil of smoke and haze rose the snow-capped Mount Rainier. It was a city at odds with itself—the battle between nature and industry.

Just then, a small, sleek head popped up beside the breakwater. We watched the harbor seal for a few minutes as the sky darkened overhead. Minutes later, large raindrops smacked against the concrete path and Steve and I ran toward our respective cars, agreeing to reconvene at a nearby pub. It was there that Steve recounted the Battle of Ballard.

IN THE EARLY 1980s, Steve was working as the state's marine mammal biologist when he was called in to help address a growing concern at the Ballard Locks. A particularly beefy California sea lion (which are nearly the same size as gray seals, with males weighing up to seven hundred pounds) had started showing up at the locks during the winter migration of steelhead trout. A local fisherman reportedly said at the time that the sea lion looked like his pal Herschel down at the dock (possibly a dig at Herschel, or the sea lion, hard to say) and the name stuck.

Herschel appeared quite oblivious to the human and boat traffic that frequented the locks, and made himself right at home at the base of the fish ladder, gulping down an impressive number of fish. After a year or two, a few more "Herschels" joined in. Observers were hired to stand watch by the fish ladder and record the sea lions' behavior—how much time they spent at the locks, how many fish each animal consumed, and whether it was the same sea lions returning each time.

Since marine mammals were federally protected, the only course of action for federal and state officials was to find a nonlethal way to keep the sea lions away from the locks. Steve knew the task wouldn't be easy. Pinnipeds are highly intelligent creatures, and the Ballard Locks had become a choice feeding station.

"You know that scene in *Jurassic Park* with the velociraptors?" Steve asked me, as we sat across from each other at a booth in the pub. I remembered it well—two kids are hiding in a kitchen with the door closed, while a raptor stands just outside, looking at the handle. Eventually, it figures out how to turn the handle to open the door. "These animals are smart," he said. "It's all about the reward they're seeking."

Steve and his fellow wildlife officials began their assignment by testing waterproof firecrackers known as "seal bombs" to deter the sea lions. For years, fishermen had used these startle devices to keep seals and sea lions away from their nets and gear. Steve would light the fuse and hurl the seal bombs, which were about the size of his index finger, into the water from a nearby platform. Others would throw them from boats. At first, they seemed to work; the sea lions stayed away from the locks. But a few days later, the animals were back.

The sea lions were quick to figure out that the people who threw the firecrackers dressed in bright orange work suits—required attire for wildlife officials working on the water—so they simply avoided the area when they spotted any orange-clad humans. "As soon as they saw the lighter, they knew what we were going to do," said Steve. When he pulled the lighter from his pocket, the sea lions would jump into the water and swim twenty yards away—about the exact distance Steve could throw the seal bomb. After the explosion, they'd quickly return to the ladder.

The team then set up acoustic deterrents—high-decibel devices that operate like underwater airhorns. The devices are thought to be quite painful for marine mammals (so painful, in fact, that those particular devices are no longer used in wildlife management). But it wasn't long before the sea lions learned to swim with their heads out of the water when they entered the noise blast zone.

Another idea was to play recordings of killer whale vocalizations. Puget Sound was frequented by two populations of killer whales, one of which preyed on marine mammals, including sea lions and seals. If the sea lions believed there were predators nearby, they'd presumably head in the opposite direction. But when they played the recordings of the killer whales, the sea lions began swimming *toward* the sounds. As it turned out, they'd inadvertently played the vocalizations of the wrong population—instead of the killer whales that preyed on sea lions, they'd played vocalizations of the killer whales that feed exclusively on chinook salmon. The sea lions were likely hoping to nab some leftovers.

It was a comedy of errors, playing out right in the middle of the city. Residents and visitors were delighted by the free entertainment, and the sea lions soon amassed a devoted fan club. People bought sea lion T-shirts, kept tallies of the score (Officials: 0, Sea lions: 6), broadcasted the updates on local radio programs. The public was openly rooting for the wily animals that seemed to be outsmarting wildlife officials.

But the fisheries and marine mammal officials trying to address a complex situation were less than enthused by the growing attention. While Steve was the point person for the state, the entire effort was being overseen by the federal fisheries service. Joe Scordino was the deputy director of the Northwest region of the National Marine Fisheries Service, and the point person for the feds. "It was challenging," said Joe, when I spoke to him later on the phone. "Because on the one hand, we had people laughing at the locks. And on the other hand, we had people concerned about the resource, essentially calling us a bunch of idiots." Unlike the spectators rooting for the sea lions to continue outsmarting the humans, this latter group was frustrated that the officials couldn't get rid of the sea lions in a more expedient fashion.

Joe, Steve, and their teams had tracked which animals were consuming the most fish and therefore had the biggest impact on the steelhead runs. Unlike the situation on the East Coast, where there was a whole host of variables to consider when it came to fish declines, wildlife officials on the West Coast perceived pinniped predation to be the primary cause for the large-scale, rapid decline of steelhead. "The only difference from earlier years was the presence of sea lions," said Joe. And while there were quite a few sea lions in the area, they determined, based on close observation of individual sea lion behavior at the fish ladder, that only half a dozen animals were responsible for 90 percent of the fish removal. Six hungry, clever sea lions wiping out entire generations of fish.

Meanwhile, the situation was getting increasingly dire. There were other steelhead trout runs in Washington, Oregon, and California, but this was the only urban steelhead run in the world. Steelhead was also the state fish of Washington and a traditional food resource for tribes in the region. But the run at the Ballard Locks was declining at an alarming rate.

As the situation at the locks ramped up, Joe began fielding calls from congressional offices across the country who had been paying very close attention. It had become clear that this battle between marine mammals and commercially important fish might have implications for the future of the Marine Mammal Protection Act. "We were kind of proving that the law needed to be changed," said Joe. The congressional offices offered more and more money to address the problem. "I told them that money isn't the issue here. The issue here is we've got to get rid of these animals and the law doesn't let us do it."

One politician offered to fund the creation of a killer whale decoy, but Joe explained that there was no way a group of highly intelligent, and highly motivated, sea lions would be fooled by a decoy. But a few Seattleites took it upon themselves to test the strategy anyway. They bought a life-sized fiberglass killer whale from a fish farmer in Scotland who had attempted to use a similar decoy to scare seals away from his fish pens. "Fake Willy" was shipped across the Atlantic Ocean, then driven cross-country on a flatbed truck. A Seattle radio station blasted rock and roll music as the fiberglass killer whale was raised and lowered into the water outside the ship canal. Steve described it as "a black-and-white salamander with a dorsal fin." Just as Joe had predicted, the sea lions looked at it, barked a few times, then continued swimming toward the locks.

Some local politicians were becoming increasingly concerned about the status of the fish. In addition to steelhead, multiple species of salmon passed through the locks. And while the sea lions migrated south to their breeding grounds off the California coast during part of the spring and summer months, Pacific harbor seals lived in the area year-round. What would stop them from feeding on the fish, especially as their populations continued to expand? They were also concerned that if salmon and steelhead were added to the endangered species list, the restrictions associated with that would negatively impact farmers, loggers, and anyone else working the landscape near the geographic zones.

At the same time, Joe was receiving calls from politicians with constituents on the animal rights side, and not just on the West Coast. "There were many on the East Coast watching what was going on," he said. "They didn't want to see us change the law to allow for the killing of [sea lions] because they didn't want any killing of seals, especially with the gray seal problems on the East Coast . . . They were very protective."

There were plenty of marine mammal defenders on the West Coast as well, and for them, the idea of killing these creatures because of a scenario that humans created was beyond hypocritical. We had destroyed the ecosystem, and the sea lions were simply doing what sea lions do: eating fish. Yet now they might be killed for doing just that.

But in Steve's view, the challenge with the Marine Mammal Protection Act, as it was written, is that it didn't account for what happens when marine mammal populations recover. In the case of the Endangered Species Act, when wildlife like bears, moose, or deer, for example, reach a certain population threshold considered to be stable, they are "de-listed," which effectively means states have the authority to impose controlled hunts. But the Marine Mammal Protection Act allows for no such option. "If I would have had a say, I would have named it the Marine Mammal Management Act," said Steve.

At the same time, managing species that spend half of their lives underwater, as opposed to terrestrial species that wildlife managers can more easily monitor, presents a whole separate challenge. As Steve had pointed out, there was so much happening beneath the surface.

Eventually, officials at the Ballard Locks attempted to physically remove the sea lions using a method Steve called "trap and haul." They captured the sea lions in cages and drove them from the Ballard Locks as far south in Washington as they could go, but the pinnipeds just swam right back to the locks. They then decided to transport the animals even further. Since the California Coastal Commission wouldn't allow the sea lions to be released in California state waters, Joe's office authorized Washington state officials to move the animals to federal waters instead. Over California's objections, Steve and his team loaded the six sea lion offenders into horse trailers and drove them more than twenty hours south along Interstate 5 to Santa Barbara, where several live news crews eagerly awaited their arrival. Steve shared a photo from the journey that shows the caravan of sea lion trailers, flanked by law enforcement, driving down a freeway lined with palm trees.

Steve and his crew then loaded the sea lions onto a boat and brought them out to Channel Islands National Park, over twenty miles off the coast of California and just a short distance from their breeding grounds. More than one thousand miles separated Southern California and Seattle, but it wasn't enough. Within a month, four of the six sea lions were back at the Ballard Locks. "It's laughable now," said Joe. "At the time, it wasn't."

While the wildlife officials continued testing various deterrent tactics, some enthusiastic members of the public began trialing their own methods. Someone put a "jiggly man" on a nearby dock—the kind used at car dealerships—to scare away the sea lions. But instead of being frightened, a few of the animals began hauling out on the dock right next to it, warming themselves beside the heat-generating box motor that blows air into the nylon balloon.

One person suggested bringing in a great white shark to eat the sea lions. Joe flatly rejected the more ludicrous ideas. "It was like, give me a break," he said. "But I had to write all of this up every time one of these ideas came in. It was very frustrating." Officials tried barrier nets positioned further downstream from the fishway, which merely shifted sea lion predation to the nets instead of the ladder. They tried shooting rubber bullets at the pinnipeds with crossbows. Nothing worked.

As these efforts were going on, a group of animal rights activists began appearing at the locks, determined to prevent the removal of the sea lions. In one incident, an activist chained himself with a bike lock to one of the cages intended to capture sea lions. But as the situation progressed, the altercations became more intense. According to Steve, some of the most extreme activists began to threaten state and federal wildlife officials, as well as members of the Army Corps of Engineers who were operating the locks but had no involvement in pinniped removal. Officials began wearing bulletproof vests as a safety precaution.

Finally, in 1994, after Joe and Steve had exhausted every nonlethal method they could think of to deter sea lions from the locks, Congress issued an amendment to the Marine Mammal Protection

Act. Under the terms of the amendment, NOAA Fisheries could authorize the lethal removal of sea lions in a highly regulated way.

There was considerable political interest and concern over any potential abuse of the new legislation. The only way to secure its passage was to require a massive amount of documentation to demonstrate that an individual animal had a significant negative impact. The legislation also required the input and approval of an independent oversight task force comprised of representatives from animal rights groups, fishermen, scientists, and conservationists. The regulations were so strict that both state and federal resource managers found them to be offensive. As Joe saw it, the implication was that wildlife officials were overly eager to kill marine mammals, when, in his case, it was a final resort option to save the steelhead. Steve had a similar perspective. "I didn't get into this business to be killing seals and sea lions," he said. "My whole career was spent studying these animals. They're fascinating." The public seemed to divide along the extremes—with some advocating for killing all of the sea lions at the locks, and others urging officials not to kill a single animal. "But they don't know the middle. They don't know what's going on in the middle."

It wasn't until two years later, in 1996, that the state of Washington was granted lethal removal authorization under the terms of the new amendment. After nearly fifteen years of testing nonlethal deterrents at the locks, they had obtained a license to kill.

Finally, they were ready to move forward. "We had the animals, we had the drugs, we had a vet on the line," said Steve. "We were going to lethally inject if no one else stepped up."

Joe was all too aware of how politically charged the situation was. He moved with extreme caution—careful to overcommunicate every decision. He was eager to make it clear across all levels of government that they were preparing to exercise their lethal authority. He had been fielding numerous calls from the Department of Commerce, the same department that had approved the permit to begin with, asking him whether their own agency had really granted authority to kill sea lions. Joe believed the calls were in response to a legal push by the Humane Society to halt the process, which had prompted government officials in Washington, D.C., to claim they knew nothing about the decision. By then, the Department of Justice was also heavily involved, preparing for a lawsuit filed by animal activist groups.

Joe had been arriving at his office around five o'clock each morning in the weeks leading up to the lethal removal date to manage an overwhelming amount of paperwork. Several federal agencies, including the Department of Justice, were heavily involved at that point. They knew that as soon as they took action, they were going to get sued.

On the morning of the scheduled removal, Joe arrived at his office before the sun was up. He'd barely removed his coat when the phone rang. On the other end, to his surprise, was Leon Panetta, President Bill Clinton's chief of staff. "It was a very short call," Joe recounted. "I was told, 'This is the White House. Stop everything. Don't do *anything* to those sea lions. We'll get back to you.' That was the end of it."

As they later discovered, Vice President Al Gore had gotten wind of the news and had managed to pull a few strings at SeaWorld. While Joe and his office had already reached out to zoos and aquariums across the country, including SeaWorld, to see if they'd be willing to adopt a few aggressive, elderly, testosterone-fueled sea lions on death row, all of them (quite understandably) had declined. But after a call from the vice president, SeaWorld appeared to have had a change of heart. The sea lions had received a vice-presidential pardon. Their death sentences were commuted to life imprisonment.

Not long after that, in early April of 1996, the federal government issued a press release with the headline: "National Marine Fisheries Service OK's Permanent Home for Seattle's Sea Lions at SeaWorld in Florida." The animals were FedExed, in an aircraft that had been set up to accommodate marine mammals, from a holding facility at Point Defiance in Tacoma—not far from where I had first met Steve—to Orlando.

It may sound like a happy ending, but it was far from it. The sea lions didn't adjust well to their new lives in captivity. The most voracious (and thus infamous) of the group died from an infection months after his arrival at Sea World, and the others died not long after that. As for the steelhead, by the time the federal approvals were finally granted, it was too late to save the fish run. The steelhead population that migrated through the Ballard Locks had been destroyed.

Meanwhile, similar challenges were playing out further south on the Columbia River, along the border of Washington and Oregon, where sea lions were gathering to prey on salmon. Between 2008 and 2023, wildlife officials have removed over four hundred California and Steller sea lions from the Bonneville Dam and Willamette Falls Dam on the Columbia River. The vast majority of these animals were killed by lethal injection, and a handful were placed in zoos and aquariums.

In recent years, the situation has begun to heat up once again at the Ballard Locks. Only this time, the salmon culprits are harbor seals.

THE DAY AFTER MY visit with Steve, I walked down to the Ballard Locks to check out the fish ladder for myself. Beneath the ladder is a cave-like underwater viewing gallery, where visitors can watch the fish as they make their way to and from their spawning grounds. It was too early in the season for the salmon runs, but

I did spot a few disoriented-looking smolts wriggling against the current.

It seemed to me that the challenge for the fish wasn't seals and sea lions, it was the existence of the dams, locks, canals, and fish ladders that impeded their migrations to begin with. Humans have condensed a process that occurs over one thousand miles of natural river and estuary systems into a mere twenty-foot passageway. With all of those fish squeezed together in an easy-to-access location, why wouldn't sea lions and harbor seals take advantage of that? I imagined humans would behave in the same way if given the opportunity.

As I approached the visitor's center, I passed a life-sized sculpture of a killer whale. It took me a few minutes to realize the sculpture was actually Fake Willy—the decoy that had been used try to scare sea lions away from the locks. It did, as Steve had attested, look a little bit like a salamander.

I stopped into the center to ask one of the Army Corps of Engineers employees how often she sees seals and sea lions at the locks. "Oh, all the time," she said. Harbor seals were more frequent visitors, as they stick around all year, she explained, whereas the sea lions migrate south to their breeding colonies in California during the summer. The seals would hang out in the pools on the ladder, lining up on the first step as they pick off the salmon, then sliding down the ladder—dinner and a ride. "It's quite a shock for the salmon," she said. "They're jumping trying to get in. But then, 'Holy shiitake mushrooms, there's a seal here.'"

I asked for her thoughts on some of the efforts to control predation at the locks. She acknowledged that the Marine Mammal Protection Act isn't perfect in that it doesn't account for high numbers of marine mammals, but she didn't believe it was fair to blame fish declines just on pinniped predation. There was also pollution, warming waters, overfishing, and fish hatcheries releasing fish into wild ecosystems. "It's a perfect storm," she said. Visitors often commented on how many fewer salmon there are now compared to when they were last here. "But then I ask them where they're planning to go for dinner, and what they'll be eating." One of the things tourists love about Seattle is the seafood, especially the wild salmon.

It was a useful reminder that we all play a role in this. We're contributing to the demand that powers the industry.

After spending hours exploring the locks, my rumbling stomach alerted me I was long overdue for lunch, so I stopped in at a nearby café for a bite. After a quick scan of the menu, I ordered a coffee, a cup of potato and leek soup, and a side of smoked salmon. It wasn't until the bright pink fish was placed in front of me that I realized my own hypocrisy.

10

The Fish Wars

ON A MIDSUMMER MORNING in 2018, off the coast of Victoria, British Columbia, an orca mother known as Tahlequah, or "J35" to scientists, gave birth to a calf. It was the first calf known to have been born to the endangered population of Southern Resident killer whales in three years. Thirty minutes later, the newborn orca was dead.

Over the next seventeen days, Tahlequah balanced the lifeless body of her calf on her nose as she carried it over one thousand miles through the water. When her calf fell, she dove down deep to retrieve it, pushing it to the surface again and again, refusing to abandon her offspring. While it's not uncommon for killer whales to carry deceased calves for a few hours, or even a day, in a presumed act of mourning, it was the first time any orca had done so for so long or had traveled so far.

Tahlequah's mournful journey made news headlines around the world—a dramatic symbol of the plight of this population of killer whales. Washington's governor appointed a killer whale task force to create a recovery plan for the population. While it was clear the orcas faced numerous threats, including boat traffic and marine pollutants, scientists believed there was one factor above all else contributing to the population's demise: the animals were starving.

While killer whales, which aren't actually whales, but rather the world's largest dolphin, are faring well as a global species, the health of individual populations can vary considerably given their highly specialized feeding habits, social groups, and even languages. In the Pacific Northwest, the transient population of killer whales, which are opportunistic hunters that feed on a wide variety of fish, marine mammals, and seabirds, were thriving, but the southern resident killer whales, which feed almost exclusively on chinook salmon, had dwindled to less than seventy-five animals. Chinook was once a widely available prey resource, but the stock in Puget Sound has declined substantially. Even after implementing various conservation measures, the population showed little evidence of recovery.

Less than two months after the death of Tahlequah's calf, another orca in her same pod also died of malnutrition. With the eyes of the world on the plight of these orcas, there was increasing pressure to bring back the salmon.

IN EARLY APRIL, A week or so before my trip to Seattle to visit the Ballard battleground, I was three coffees deep into research on West Coast pinniped science when I happened across an article published by KNKX, Tacoma's NPR station. The article, penned by journalist Bellamy Pailthorp, focused on the frustrations of coastal fishing tribes in the state over the growing numbers of seals and sea lions preying on salmon. I was particularly intrigued by the perspectives of a tribal elder named Ramona Bennett, who advocated for a tribal seal hunt to protect the salmon while also bringing back a traditional menu item for the tribes. There was just one thing standing in her way: a piece of federal legislation known as the Marine Mammal Protection Act. On this point, she was quoted as saying, "And let them arrest us. We'll never resolve this through legislation, it'll have to be through litigation, which is clearer and quicker."

I studied the photograph of Ramona included in the article, finding myself strangely drawn to her. A small, thin woman, she was seated at a picnic table, dressed in a long-sleeved black T-shirt with large silver hoop earrings, her gray hair pulled back, a cigarette balanced between her fingers. She was smiling, with a glimmer in her eyes as though she'd just made an irreverent joke. Who was this woman?

A short Google search later, and I was captivated. In her mid-eighties, Ramona was a tribal elder and former councilwoman for the Puyallup (pronounced "pew-ALL-up") Tribe in what is now Tacoma, Washington. A well-known human rights activist, Ramona was one of the leaders behind a series of protests in the 1970s to defend Native fishing rights, an event that has come to be known as the "Fish Wars."

Eventually, I tracked down a possible phone number for Ramona, although it seemed an unlikely match given that I'd found it on a flyer listing contact information for the Puyallup Fireworks Commission. I couldn't imagine a woman in her eighties being on a fireworks commission, but I was hoping whoever answered might be able to connect me to her, so I dialed the number. To my surprise, the slightly raspy voice that answered belonged to Ramona Bennett herself.

Since I hadn't expected to actually reach her, I began stumbling through an ill-prepared explanation of why I was calling. On a whim, I asked if she'd be open to meeting in person when I was in Seattle the following week. Instead of responding, she asked if anyone was taking me out on the water to look at the seals. They're all over the big water, she said, following the salmon like "packs of wild dogs." The salmon were at risk for a whole host of reasons, she added, from chemicals poisoning the estuaries to industry heating up the waters. But the few salmon that do survive are then eaten by the seals and sea lions. They used to stay out in Commencement Bay or in the Salish Sea, but now they're seeing seals foraging all the way up the Puyallup River.

"No one gives a rat's ass about Indian fisher people," said Ramona. "But people out here are concerned about the orcas." Since the tribes couldn't hunt seals because of the Marine Mammal Protection Act, she suggested PETA should have them spayed and neutered. They could clip their ears, she said, the way they do for feral cats, to indicate which seals had been fixed. "Well anyway, seals don't have ears," she added, laughing. "So we can't figure out what to clip."

She told me to stand by while she tried to find someone to take me to see the seals. "If I forget, you call again."

A couple of hours later, Ramona sent me a message: *"Alix plz txt Jim Jim Rideout. He's a Puyallup councilman and fisherman. He's committed to helping reduce the problem of too many seals."* She included his phone number. I thanked her and quickly reached out to James "Jim Jim" Rideout, who got back to me later that evening.

"I know when Ramona calls, I better clear my schedule," he said. "Doesn't matter if I'm on my way to the White House." James was a fisherman as well as one of seven elected members of the Puyallup Tribal Council. He had also been the fishing partner and close friend of Ramona's son, Eric Bennett, who died from cancer in 2014, and had been striving to highlight the issue of pinniped predation on salmon in the final years of his life. James told me that my focus on the impacts of increasing numbers of seals, and my desire to understand different perspectives on the issue, had struck a nerve for Ramona.

If it weren't for people like Ramona who fought for our land

and rights, he said, there wouldn't even be a Puyallup Tribe today. "I might be sitting at the council table where Ramona once sat, but she built that table."

James offered to wrangle up a boat while I was in town. He wanted me to see with my own eyes the battles the tribe continues to fight—not just with seals and sea lions, but also with industry and pollution. The only challenge was choosing the right day for it. There's a season for salmon, he explained. You watch the flowers to know when the fish are running; the daffodils will tell you the salmon are in. It was all about timing.

THE RIGHT TIME ARRIVED on a surprisingly sunny spring morning, when I met James at the Chinook Landing Marina in Tacoma. As we walked down the dock to the boat, I eyed James's black fleece hat, which had a skull and crossbones printed on the front—except the crossbones weren't actually bones, but rather, bone-shaped salmon, with the words "Spawn Till You Die" printed underneath.

James, in his midfifties, and his son, Jake, thirty, had been in the midst of transferring crab pots from their fishing boat, the *Gloria Jean*, to the small motorboat that James and I would be heading out in that day. After we finished our seal survey, he said, he'd need to set the traps so he could provide Dungeness crabs for a tribal funeral the following day.

I stepped onto the skiff to help load the crab pots. The Dungeness crab fishery was closed at the time—crab numbers had plummeted for reasons unknown to fisheries scientists (although some suspected the declines had something to do with "The Blob"—a giant patch of warm water in the Pacific Ocean that had been wreaking havoc on weather systems and marine ecosystems from the Gulf of Alaska to Baja California, Mexico). But the tribes had a special ceremonial fishing allowance, as crab and other shellfish were traditional food resources. As soon as we'd loaded up, we waved goodbye to Jake and motored out of the marina.

James explained that they hadn't been able to get out on the water to fish for the past week after a factory trawler docked nearby had caught fire. As we headed for Commencement Bay, we passed alongside the burnt ship. The massive vessel, which looked like it had been barbequed, was listing hard to one side, still docked near the entrance to the bay.

The trawler was owned by Trident Seafoods, the largest seafood company in the U.S. While it was unclear what had caused the fire, it was the second ship fire for the company in less than three years, in the exact same location. The ship had burned for six days, prompting a multiday shelter-in-place order for nearby residents due to air quality concerns. Meanwhile, the U.S. Coast Guard halted boat traffic in the Port of Tacoma while firefighters attacked the flames.

One of the most urgent concerns was preventing the flames from reaching the nearly ten tons of Freon tanks aboard the ship. This chemical coolant is used to freeze fish during long journeys offshore, but if it's accidentally released into the atmosphere, it can wreak havoc on the Earth's ozone layer. Yet by the time the firefighters had battled back the flames enough to access the tanks, they were already empty. The damage was done.

During this time, Puyallup fishermen, including James and Jake, were blocked from getting their boats out on the water to harvest. But one of the most shocking aspects of this, for me at least, was that despite the intensity of the fire and its devastating environmental impacts, the event had received almost no news coverage, even locally. Yet here was this giant ship, burnt to a crisp—a symbol of just one of the harmful impacts of industrial offshore factory fishing staring us right in the face. A literal black mark on the landscape. James smiled at my incredulity, as though he'd been expecting it. "Oh, just you wait," he said.

IN THEIR NATIVE LUSHOOTSEED language, the Puyallup are known as the *puyaləpabš*, which translates to "people from the bend at the bottom of the river." The Puyallup River Delta was once a pristine habitat for salmon, marine mammals, shorebirds, and other wildlife. But in the nineteenth century, the arrival of European settlers kicked off the long decline of these species as Tacoma was transformed into an industrial city—tidal flats and wetlands, once thriving ecosystems, were demolished to make way for lumber mills. The meandering rivers and streams of the river delta were filled and dredged and reshaped into artificial waterways that better served the economic needs of the city. By the 1980s, Commencement Bay, which sits at the mouth of the Puyallup River and is home to the Port of Tacoma—one of the largest and busiest container ports in the country—was listed as one of the most polluted bodies of water in the nation.

James steered the skiff toward the Puyallup River, the main source of freshwater that flows into Commencement Bay and the Port of Tacoma. At the mouth of the river, a large pulp and paper mill puffed out plumes of white smoke. The mill, which had been in operation for nearly a century, was the long-presumed source of the "Aroma of Tacoma," as it had emitted a putrid scent of sulfur for decades, until new technologies were installed in the early 2000s. (The smell was so strong that during a 1985 concert tour in Tacoma, Bruce Springsteen became sick and had to leave town early.)

I asked James about the salmon they fished for in the river, if there were ever issues with biotoxin levels or contaminants that prevented them from being able to harvest. He explained that all of the species they fished for, including salmon and Pacific geoduck (large saltwater clams), were tested at the University of Washington as part of their quality and health inspections. They hadn't had any issues with toxicity levels, but James harbored his own concerns about the impacts of pollution. "You tell me why my mom died of cancer, why Eric died of cancer," he said, sadly. "And I'm also a cancer recipient. We've lost family members from living this way of life. The cause and effect of what's happened with industry and in our waters."

Past the mill, we continued a short distance upriver, where rows of brightly colored shipping containers, sandwiched between drab warehouses and semitrucks, lined the riverbanks. Suddenly, we noticed a splash about fifteen feet from the boat. James cut the motor. Sure enough, a familiar-looking, round, glistening head emerged, and then another—harbor seals. The seals turned to look at us briefly before disappearing beneath the water, then quickly reappeared alongside the riverbank. Soon after, several other seals popped up on the other side of the boat. I counted at least half a dozen seals swimming and splashing beside us.

Caught up in my own amusement over the seal sightings, I nearly forgot why I was there, until I looked back at James, who was far from amused. The expression on his face didn't seem to be one of anger, or even frustration. It appeared to be sadness. While we couldn't see what, or even if, they were eating (harbor seals often consume their food underwater), James had no doubt in his mind that they were feeding on salmon.

"It's not even our fishing season yet," he said, shaking his head. "You can see it for yourself."

James had grown up salmon fishing on the river, right in this very spot. The tribe had been working to preserve the salmon, a critical part of their culture and identity as "salmon people."

One of their efforts was to raise salmon in a nearby hatchery, then transport the fish by truck to a release site upriver. It was a method also being used by Washington State fisheries officials and other tribes to boost salmon populations, and had received significant federal support, but it came with its own set of challenges. Numerous studies have shown that hatchery-grown salmon—long touted to be the salvation of salmon populations in the West—can weaken the diversity and health of wild stocks. But alternative options to rebuild the populations, such as halting the fishery or dismantling the dams that power irrigation and electricity, were far less appealing.

There was another issue as well—huge numbers of hatchery-grown salmon don't survive, for a whole host of reasons that include pinniped predation. James explained that as soon as they released the young salmon, the seals would appear where we saw them now, large numbers of them lined up across the mouth of the river to feed on the salmon as they made their way into the bay from the hatcheries. It suddenly made sense what Ramona had said to me on our call days earlier: "We're raising seal food."

James's grandmother had also fished. In her later years, she watched her grandson from the riverbank in her wheelchair. "We were out here as a family," he said, as he eyed the seals. "I'd be on the river in a diaper with an RC Cola, bologna sandwich, and I was happy." He was silent for a moment, staring at the riverbank as the skiff gently rolled in the current.

WHEN THE TIDE TURNED, we motored from the river to a nearby marina so James could pick up fuel for the boat. He tied the lines at the dock and we walked up to the store for fuel and coffees. As I poked around the shop, James recommended a pair of socks as a gift for Ramona. "She'll love you for these." The socks featured an image of a bald eagle holding a salmon in its talons, with the words, "Fish Assassin."

As we paid for the socks and coffee, James asked the young man at the register for his thoughts on the impacts of seals and sea lions. The store clerk looked at us for a moment. "I've heard stories of some of these guys that hit them with their boats on purpose for eating the fish," he said. "I'm not sure I agree with that. I mean, they gotta eat, too." As we walked out, James looked at me and shook his head. "He's too young to understand," he said, quietly. "He hasn't seen the impact."

But the young clerk wasn't wrong about the intentional boat strikes. Days earlier, NOAA's Office of Law Enforcement had posted an alarming video of a boat on the Columbia River in Oregon chasing after sea lions at top speed. Dozens of animals dove beneath the water to escape the vessel as it tore from one group to the next, slamming across its own wake as it motored in figure eights. NOAA was actively seeking information from the public about who might be responsible.

Back on the dock, James was untying the lines when I noticed a sport fisherman in a neighboring skiff suspiciously eyeing the crab pots. "You crabbing out there?" he asked, pointedly. "These are Puyallup Tribe ceremonial crab pots," James calmly replied, continuing his work. "They're for a funeral." "Nice," said the fisherman. "You're lucky." James paused mid-task and looked hard at him. "They're for a funeral," he repeated, slowly. "I've had a *loss*." "Well, nice and unlucky," the fishermen said quickly. "I just mean I want to be able to get out there crabbing, get some for myself to eat."

"This lady here is writing about seals and sea lions," James said, resuming his task. "What do you think of them?" "Get rid of them all," the fisherman said immediately, turning to me. "There are too many. If they want to protect these animals, fine. But give me a number and then get rid of everything else. The numbers just keep growing. They eat everything out there."

THE AIR WAS CRISP as we motored across the bay, and I was grateful for the warm coffee in my hands as I gazed up at the sun. After James set the crab pots, which he'd baited with salmon and geoduck (not bad for a final meal), we returned to the Puyallup River to retrieve his truck, which he'd parked beside the riverbank.

As we were driving back to the marina, I asked James what he thought the ideal situation with seals and sea lions would be. What did success look like to him? He considered my question for a moment. "With all this effort that we go through," he said, "success would be to thin out that population, and to be able to see our salmon numbers come back. If we wait for the state to do something, it's never going to happen, right?"

"But how would you determine how many animals to kill each year so that you're seeing an impact, but aren't killing more than you need to?" I asked. James slowed the truck as we approached a light, then looked at me, thoughtfully. "Well, that's a really good question."

AFTER RETRIEVING MY RENTAL car from the marina, I drove along a busy road that runs parallel to the Puyallup River until I spotted the sign for Ill Eagle Fireworks. The sign featured an amusing illustration of an eagle coughing. I parked in front of one of several mobile homes on the property, hoping it was the right one. Just then, the front door opened and a man emerged in a motorized wheelchair. He sped down the ramp at a surprisingly fast clip, with his large dog—an Akita that must have been upward of 80 pounds—in close pursuit. I stepped out of the car as the man reached the end of the ramp. He then stood up from his wheelchair and smiled warmly at me. "I'm Clyde," he said, extending his hand. "And this here is Sammy." He motioned toward the dog, whose tail was wagging furiously as it barked and ran circles around us.

Clyde Bill was Ramona's husband. In his seventies, Clyde was a tall man with a warm smile and a presence I found immediately comforting. "I'm Alix," I said, laughing at Sammy's enthusiasm as I scratched the fur behind her ears. "Thank you so much for having me." "Go right inside and make yourself at home," he said. "Ramona will be right out."

I walked up the front steps as Clyde and Sammy disappeared behind the house. "Hello?" I called out as I walked inside. No one responded. I made my way through the living room toward the kitchen. Finding it empty, I took a seat at the dining table. The rooms were filled with artwork, pottery, piles of papers, sculptures—a home bursting with life, culture, family. The Puyallup Tribe's salmon emblem was everywhere.

As I waited, I reached down to pat one of several cats milling about, as it gently rubbed its head on my ankle. A few minutes later, Clyde appeared at the door. "Ramona's in her bedroom. You can come on in." I followed him into the room, where Ramona was seated on the edge of the bed. She was facing the window while fastening one of her earrings, which she later told me were South Dakota quillwork medicine wheels—a sacred symbol. "Have a seat there," said Clyde, pointing to a chair beside the bed.

"That's the interrogation chair," said Ramona, as she turned toward me, the same glimmer in her eyes that I'd seen in the photograph. "Then this is the chair for me," I said, smiling as I sat down. "It's so nice to meet you, Ramona." As she eased herself back on the bed, Clyde rushed over to assist her. He then stepped out of the room to leave us to chat.

Almost immediately, one of the cats jumped onto the bed to sit with Ramona. She explained that the cat had been dropped off by a girl they'd raised who had moved to an apartment that didn't allow cats. Their dog, too, had originally belonged to her grandson. "When you're getting ready to go away to college, you're supposed to get a big dog and dump it on your grandmother or your mom, whoever is closest," she said. "In this case, his mom and I both lived here. But she's really been perfect."

At that moment, her phone rang. Someone was outside looking to buy fireworks, so Clyde headed out to broker the sale. And Ramona settled in to tell me about the Fish Wars.

BACK IN THE 1850s, Native tribes in the West, including the Puyallup, signed treaties with the U.S. government, which effectively ceded their land but secured their rights to continue fishing in their ancestral waters. Between the early 1900s and the 1970s, clashes between Native and non-Native fishermen, as well as state officials, were on the rise. Eventually, the states of Washington and Oregon passed laws to limit Native people's rights to fish for salmon—laws that were in direct violation of their federally protected rights that had been enshrined in those treaties.

In the summer of 1970, Ramona and a group of activists set up a fishing camp on the banks of the Puyallup River. The camp was positioned as a "fish-in," similar to the "sit-ins" that had been taking place to protest the Vietnam War. After a few days, the camp began to expand, with tribal members and activists joining from other regions to stand in solidarity with the Puyallup people. Ramona was a member of the Puyallup Tribal Council at the time, and one of the leaders helping to organize and manage the camp.

"It was the tail end of summer," she said. "We'd been there for six weeks. And we were so settled in I was fearful that we were going to be stuck there for the whole damn winter. But as soon as the weather got bad, we sent all of the kids out of the camp. And that's when they came in on us."

At first, it had appeared state officials didn't know what to do with the fishing camp. "We were a peaceful camp on our own property, minding our own damn business. They had absolutely no reason to fear us or hate us. We weren't doing anything to anybody." But eventually, the chief of police told Ramona that they needed to cease and desist. "I said, 'What, living?' We weren't bothering anyone." But the chief had orders to break up the camp, and told Ramona that if they didn't disperse, the officers would come in and remove them by force. "And I said, 'If you come on this property, you will be trespassing, and you don't have jurisdiction.'"

On September 9, 1970, the state raided the camp, sending in armed officials to break it up, but tribal members, some of whom were armed themselves, refused to leave. The confrontation became violent, and more than sixty people were arrested, according to the *Seattle Times*, including Ramona.

"It was a war," said Ramona. "They sent five hundred and fifty pigs with flak jackets and rifles and gas canisters, and they attacked us. People sat down to eat their evening meal and they thought they were watching John Wayne theater because Indians were being clubbed and thrown around and dragged. Washington State is attacking these Indians, and what are the Indians doing? They're just fishing. Why are they fishing? Well, they have a treaty right. This had been going on for a hundred years—usually in the upper river, usually after dark, but they attacked us in broad daylight."

But Ramona and the other tribal leaders had been strategic in choosing where to set up the fishing camp. They'd selected a highly visible location, which enabled easy access for members of the public, and most importantly, the press. "By then, we really knew how to use the media. I'd had drama and debate in high school, I used everything. I mean, these were real fishing families. They lived on the water. It was their whole life. So we had this battle." As the battle unfolded, photographers captured images and video footage that shocked the nation, and the world.

The federal government was in the international spotlight. On behalf of the tribes, the feds sued the state of Washington. It took four years, but in 1974, in a landmark ruling, Judge George Boldt ruled in favor of the tribes. Boldt affirmed the treaty rights of the Native fishing tribes, asserting that these rights equated to 50 percent of the fishing harvest. Boldt's ruling infuriated many non-Native commercial and sport fishermen, and received intense opposition from Washington state officials, but after years of protests and appeals, it was eventually upheld by the U.S. Supreme Court.

For years leading up to the Boldt Decision, the federal government had refused to recognize the Puyallup Tribe. "They were so confident that they had successfully disposed of us," said Ramona. "The way we were being terminated was by the rolls never being updated. The Bureau just wouldn't certify any new members. And so, I went and took the rolls away from them, and we certified our own members. Nobody cared because we weren't recognized anyway, but we kept track of who we were. But when we caused that court case to happen, suddenly we were a tribe again . . . Any Puyallup that survived descended from some very quick, clever people, because they meant to kill us all."

Ramona and Clyde's house was just a mile upriver from the exact place where the Puyallup and other Northwest fishing tribes had faced off with law enforcement. The bridge overlooking the encampment, which has since been renamed the Fishing Wars Memorial Bridge, serves as a constant reminder of what they fought for, what they have continued to fight to protect.

I SCRIBBLED FURIOUSLY IN my notebook as Ramona recounted the details, trying to record as much of what she shared with me as I could. I was overwhelmed by how little I'd known about this piece of history. When I looked up, Ramona was watching me, a mildly confused expression on her face. She then turned and reached for something on the opposite side of the bed, what looked to be a thin book with a black cover. She clutched it in her hands for a moment, looking at it, then handed it to me. "I think I will give this to you and take my chances I can get another," she said. The book turned out to be a leather-bound notebook with the Puyallup salmon emblem printed in red on the front cover.

"This is beautiful, but are you sure—" I tried to reply.

"I am sure. It's so much better than your little notebook you're carrying." There was no question the Puyallup notebook was far superior to my plain spiral-bound sale purchase from Staples.

"Now you look absolutely official."

"I have something for you as well," I said. While my gift felt entirely inadequate, I pulled the "Fish Assassin" eagle socks from my bag and handed them to Ramona. She broke into a wide grin. "Who gave these to you?" she asked. I laughed. "All right, Jim Jim may have played a role in this."

BEFORE HE PASSED AWAY, Ramona's son Eric had been trying to raise awareness about the impacts of increasing numbers of seals and sea lions on the salmon fishery. Days before my visit, Ramona and Clyde had shared a video of Eric reeling in empty nets, as sea lions swam alongside the boat.

Ramona told me that when Eric was young, he would pack a thermos of coffee and several sandwiches, then run across the road from their house and position himself near a culvert by the river. "He'd wait for a boat to come up the river, and he'd jump on," she said. As the tide came in, they'd go up the river, and when it turned, they'd come back down. The Native fishermen who gave him a ride would let Eric lay the net and pull the salmon out. "And they would tell him stories, and eat his sandwiches, and drink his coffee. And he learned everything about everybody, and the water, and the tides, and the salmon, and the runs. That's when he was ten, he did that." She laughed. "And I know, who lets their ten-year-old child go fishing with random Indians that you don't even know? But I mean, he loved it and there was no stopping him. And he took all of that information with him, and of course shared it with Jim Jim. And he lived a good life."

"I have a plan for the seals and sea lions," said Ramona. "You can't change anything through lobbying. It's too slow. And so I want us to get a team of tribes together and harvest seals and sea lions, and invite legislators and people of influence to a big seal meal. And get a bunch of hunters arrested and then go to court, because court is a lot faster.

"We are the only ones with the *right* to harvest salmons. The sport fishermen and the commercial fishermen, they have an *opportunity* and a *privilege*. But we have a right. And so the salmons that have been destroyed by the seals are ours. And so we own the seals and the sea lions." She paused, smiling, watching me. "This is my convoluted logic, but I can make it work. My high school debate never fails."

It wasn't hard to understand why Ramona wanted to take a direct approach to solving the issue of pinniped predation, and it certainly wasn't the only battle she was fighting to bring back the salmon. The Puyallup Tribe had recently filed legal challenges against Puget Sound Energy in an attempt to block a proposed liquefied natural gas facility, which would contribute to already sky-high rates of air pollution. Cancer risk from air pollution in the region is ranked among the highest in the nation. The tribe was fighting a separate legal battle against Electron Hydro over the construction of a dam that had been polluting the water and blocking upstream fish passage.

Since time immemorial, Native people have lived along the rivers and estuaries, along the coast of the Salish Sea, as had salmon, steelhead, seals, sea lions, killer whales, and an abundance of other wildlife. In the mid-1800s, when the tribes signed treaties with settlers, they preserved their rights to fish and hunt in their "usual and accustomed grounds and stations," but the U.S. has broken those treaties time and again. While the 1974 Boldt Decision affirmed their rights to 50 percent of the salmon harvest, that means nothing when there aren't any salmon.

The Marine Mammal Protection Act had been successful in its goal of bringing back overhunted populations of pinnipeds and other marine mammals, but for Ramona, at least, the federal law also presented yet another threat to their way of life.

A FEW MONTHS AFTER my visit, Ramona was listed by *Forbes* magazine as one of the 50 most impactful women over 50 in 2023. In the article, she was described as a "pioneering civil rights leader who has . . . fought for social and healthcare services and tribal fishing rights." The final line read: "In 2022, she began campaigning to amend a law that she believes undermines the fishing rights of tribes in the Pacific Northwest."

11

The Chase

IT WAS A DREARY spring day in Maine. The air was cool and damp, the early-morning fog had lifted into a gunmetal sky. As I drove to Marine Mammals of Maine, I stewed over the realization that despite all of my research, I was no closer to understanding what could actually solve the problem (or, more accurately, *problems*) at the heart of the human-seal conflict. And yet my list of questions had multiplied exponentially.

Foremost among them: how could we begin to have the conservation about how or whether to manage seals without better understanding their movements, behavior, and ecological functions? Proponents of seal culls often demanded a number from wildlife officials—the threshold at which scientists could determine the population was sustainable. Once that number was reached, seal hunting season could theoretically open. But that concept, which would be similar in theory to the way we manage terrestrial species such as deer, requires a substantial amount of data on the size and status of a given species, along with a far deeper understanding of its ecological roles. Our insights about marine species and ocean ecosystems, based on many decades of intensive research, still seemed so limited.

My West Coast journey had given me an even greater appreciation for our reliance on fish—the last wild commercial harvest that exists in the world. Yet when it came to addressing the impacts of pinniped predation on fish, I'd come to believe there was a huge difference between killing a small number of sea lions at dams, where the impact of their predation could be easily quantified, and culling seals—a blanket removal of animals to reduce the population without any evidence that doing so would bring back the fish.

"DID YOU TELL HER what Sunshine did?" Lynda asked Dominique, grinning. Dominique groaned and rolled her eyes. Stubby, one of Dominique's two dogs, was nestled beside me on the couch as I dutifully rubbed his belly. Amidst all my winter and spring travels, it was my first visit in weeks, and I was elated to hear the latest seal gossip.

In my absence, the four gray seal patients that had arrived at the center back in January had received official names. As part of a new sponsorship program to help fund the seals' care, the staff had offered four local businesses the opportunity to name the now-healthy seals that were nearing their release dates. The feisty Cape Cod seal I'd last seen flinging snow across his enclosure was now named Breaker, which seemed quite fitting. The quiet seal with a fractured flipper, formerly known as Number Two—a gentle wallflower whose soft gaze turned my belly to jelly—was now Titan. Number Six, the wayward wandering seal with a devoted fan club, was now Dexxy, and Number Four, Dexxy's energetic and sassy pool-mate, was named Sunshine.

Dominique sat beside Lynda on the couch, absentmindedly hurling a Frisbee across the room, much to her other dog, Fin's, delight, as she recounted Sunshine's recent antics. Days earlier, she had spent hours cleaning Dexxy and Sunshine's pool, hoping to capture clear underwater footage of the two seals feeding on frozen herring to share on social media.

As soon as she'd finished, she suspended a GoPro camera over the side of the pool to record the seals while she prepped fish for their upcoming feeding. Because of the team's need to limit interactions between the seals and humans as much as possible during their rehab, sharing photos and videos on social media was often the only way to keep the public informed about the animals' progress. The updates were an important way to educate people about seals, and, in the case of Dexxy, a way to proactively address the influx of inquiries prompted by his early fame. But the updates were a crucial part of the team's fundraising efforts. As the animal welfare campaigns of the sixties and seventies had proven, photos and videos of adorable seal pups are a particularly effective way to inspire donations.

But as soon as the camera appeared, Sunshine wasted no time. She swam up to the camera and gripped it in her mouth, pulling it into the pool. She then dropped it in the center of a large pile of poop she'd released mere seconds after Dominique had finished cleaning. "You can see her one eyeball staring through the poop at the camera," Lynda said, no longer able to contain her laughter. "After all that work, I basically just filmed a pile of seal poop," said Dominique.

When she wasn't messing with Dominique's video efforts, Sunshine had also been keeping Dexxy busy. As I skimmed through the weeks of feeding notes, it was clear she'd been outcompeting the celebrity seal during mealtime. The rehab feeding process varies depending on the age and health of the seal when it's admitted, but generally begins with a fish smoothie administered through a tube. This feeding method limits the amount of time the seal needs to be handled. The seals then graduate to an "assist feed," where staffers wait for them to open their mouths (often in a fit of rage, as "hangry" seals are quick to vocalize), at which point, the staffer places a stiff, half-frozen fish as far back in their mouths as they can, gently guiding it down their throats. At first, the seals by no means enjoy this, but eventually, they catch on, and begin to gulp the fish down without human assistance. Eating frozen fish doesn't come naturally to seals, but in a rehab setting, it's the only feeding option. Even so, staff have found ways to get as close as they can to replicating a more natural feeding process.

After seals have mastered the unassisted fish gulp, they're moved to a small pool filled with water, where they learn to "chase" a fish. The fish, in this scenario, is still quite dead, but the staff grip it with a pair of long tongs, which they drag back and forth through the water until the seal "catches" it. In the final step before release, the seals graduate to the big pool. Here, the staff climb a small set of stairs next to the pool, crouch behind a wall at the top so the seals don't associate the arrival of fish with humans, and hurl the fish into the pool. The seals must learn to compete with each other for the fish—mastering a skill they'll need to survive in the wild.

The feeding notes for Dexxy and Sunshine painted a clear picture of the superior fish collector: *Dexxy ate slowly compared to Sunshine, who ate far more . . . Dexxy not as fast as Sunshine . . . Dexxy ate 1 fish while Sunshine ate 4.* Sunshine became so quick with her fish consumption that she even began taking fish out of Dexxy's mouth. But after a couple of weeks, he finally began to catch on. *Dexxy still slower but competitive . . . Dexxy got a few extra when Sunshine slowed down . . . He held his own against Sunshine . . . Tried to take one from her, unsuccessful.* Meanwhile, the staff closely monitored the weight of each of the seals as they neared their release date. They were about to level up to the biggest pool of all.

THE EARLY SPRING MORNING was warm and bright as a procession of nearly 150 eager spectators, including dozens of children, marched in pairs down a narrow beach path in Phippsburg. It was the exact same beach where the staff had released two harbor seals over six months earlier. But thanks to Dexxy's celebrity status, there were at least ten times the number of spectators in attendance, along with several local news crews scrambling to set up their tripods and cameras so as to capture the seals' journey to the beach.

Choosing when to release seals back into the wild is a tricky process. The seal's health and readiness must be approved by Lynda as well as Kip Temm, a licensed veterinarian who oversees the medical care and approves all major medical decisions, working closely with Lynda. Finally, it must be approved by the National Marine Fisheries Service, to ensure the rehabbed seal doesn't present any kind of disease or other threat to the wild population. Additionally, Lynda must closely monitor the marine forecast. Even on mild, sunny days, a strong current or rough surf is enough to confuse a young seal and cause it to re-strand shortly after release.

The two seal crates were again positioned side by side on the beach near the surf, as the crowd of humans formed two walls on either side leading toward the water. After a few words from Lynda and an enthusiastic ten-second countdown from the crowd, the cage doors were opened.

Cameras clicked, the waves rolled in, the wind whistled, and the crowd was still. The seals, however, stayed put. I couldn't exactly blame them. After months of dedicated care and guaranteed meal service of Lynda and her team, abandoning the safety of their crates to pass through a tunnel of gawking humans toward a giant, roaring mystery pool, may not have been the most enticing option.

Unfortunately for Dexxy and Sunshine, Lynda was adept at managing reluctant seals. She was soon offering Dexxy some unwelcome encouragement in the form of a gentle tip of his crate. As he cautiously emerged from his kennel, a spectator leaned forward to capture a video on his phone. Behind him, a videographer for a local news station snapped, "Get out of my shot!" Dexxy, oblivious to his fame, plopped out of his crate onto the sand and lifted his head, revealing what looked like a bizarre fascinator.

Earlier that morning, back at the center, the staff had affixed a satellite tag, which resembled a small walkie-talkie with a six-inch antenna poking out, to the back of Dexxy's head with a specialized, non-toxic glue. The tag is designed to stay on for up to nine months, at which point the seal sheds it off naturally during his annual molt. But during that time, it collects a massive amount of data. The tag is programmed to ping when the seal surfaces to breathe, as it transmits the date, time, and an animal's geolocation to satellites. Tags can also measure how often seals are diving, how deep they go, and how long they spend at that depth, as well as the length of time a seal is hauled out on land.

Dexxy's tag had been provided to Lynda by Kimberly Murray, the seal research lead at the Northeast Fisheries Science Center in Woods Hole, Massachusetts, part of the National Marine Fisheries Service. The satellite tags, which cost thousands of dollars per device, are typically used to monitor wild seals captured by trained research teams, as opposed to animals that have received extended human care. But Kimberly had offered an extra tag to Lynda, and she'd selected Dexxy as the recipient. For Lynda, the tracker would provide valuable information about how a rehabbed seal fares upon release, offering insights into its behavior and movements in the wild.

After Dexxy and Sunshine had disappeared into the surf, I overheard a spectator ask one of the staffers if Dexxy had waited for Sunshine before he swam off. I smiled. It wasn't long ago that I'd asked Lynda a similar question. It was easy to assume the seals, having shared a pool for months at the rehab center, would stay together in the wild—at first as best friends (at least in my version of this Disney movie), and eventually as lifelong, monogamous mates. Of course, my visit to Sable Island had dashed any preconceptions of the happy, peaceful lives of gray seals in the wild.

But say you wanted to know what *actually* happens after a seal jets beneath the waves. Satellite tagging and other research tracking technologies offer useful insights into their underwater lives. For scientists and wildlife managers, understanding seals' geographic preferences—both on land and in the sea—is a useful way to predict where human-seal interactions are most likely to occur. While tagging provides important clues, there's still so much more scientists are hoping to learn about seals' underwater worlds.

OVER THE YEARS, RESEARCH teams have converged on seal pupping and haul out sites throughout New England to collect as much data as possible. The information is used not only to help scientists better understand the population status and ecological role of seals, but also to inform fishermen and fisheries regulators who are looking to reduce conflicts between seals and humans. Scientists have amassed quite a bit of information—from the number of pups born each year to the health and behavior of seals hauled out on the beaches during pupping season. As soon as the animals disappear underwater, however, their lives became a virtual mystery. Where do they go? What do they eat? Do they hunt on their own or in groups? When and how do they sleep?

Tracking seals' underwater movements can offer researchers insights about the lives of seals, but there are other benefits as well. For years, scientists have been placing satellite trackers on deep-diving Weddell seals and southern elephant seals in Antarctica to gather data on ocean temperatures and salinity. These data have contributed to our understanding of how the climate crisis is altering marine ecosystems. More recently, Australian scientist Clive McMahon and his colleagues compared data from more than 500,000 individual seal dives against an existing map of the seafloor. Seafloor maps are often limited at best, due to the inherent challenges in deep water ocean exploration, particularly in extreme climates. The scientists found that in a number of cases, the seals appeared to be diving far deeper than the maps indicated would be possible.

In a 2023 study published in *Nature Communications Earth and Environment*, the team revealed that data from these seal dives, combined with sonar depth mapping from an Australian research ship, had helped them to find a hidden underwater canyon in eastern Antarctica. The canyon, in Vincennes Bay, extends more than a mile beneath the surface of the water. The findings could offer crucial insights into how Antarctica's ice will melt. Because the deep ocean around Antarctica is warmer than the surface waters, these deep-seafloor canyons enable warmer water to flow to the ice along the coast. The more information scientists can gather about the presence and depth of these canyons, the better they'll be able to predict future melting.

In addition to providing ocean mapping support, seals are considered to be sentinels of ocean health. Because of their frequent travels between land and sea, they can serve as early warning signals about emerging threats, from marine pollutants to disease outbreaks, such as the 2022 outbreak of highly pathogenic avian influenza in gray and harbor seals in the Northeast.

YET ONE OF THE major challenges for seal scientists in the U.S. is how little money there is to support their research. Seal studies tend to fall to the bottom of the priority pile when it comes to government funding, lagging far behind critically endangered species like the North Atlantic right whale, or commercially important species like Atlantic salmon. It's not easy to convince a government agency to prioritize funding to study a population that appears to be thriving. Yet it's exactly because of these thriving populations that conflicts between seals and humans have been on the rise. And without research data, it's impossible to address the problem, or even to fully understand it.

"It's really hard to do research on seals," said Kimberly. "And yet because they're recovering, and they're so close to shore and interact with humans, there's all sorts of social conflicts." She often hears from people who complain about seals, arguing that they're eating all of the fish, or they're attracting great white sharks to New England beaches. "It's frustrating for me because we don't have the means to get the information needed to inform those debates with the public."

Gordon Waring, Kimberly's predecessor at the Northeast Fisheries Science Center in Woods Hole, who has since retired, echoed Kimberly's perspective. I met Gordon for coffee one morning when he happened to be driving through Maine. We met in Brunswick, not far from the rehab center. "There's so much of a human dimension with seals, more so than so many other species," he said over the din of the pre-lunch crowd. "But how do you tell this story so that people understand? Humans are so human-centric. It's us and it's them. And it's always us, and it's never them."

Gordon had worked for NOAA since 1973, first as a fisheries biologist and then as the seal team lead. He explained that the federal funding challenges with seal research had always been there. "We tried all kinds of ways to get more money," he said. "Seals just aren't a species of concern."

AFTER DEXXY'S RELEASE, HIS tag pinged his whereabouts each time his head surfaced above the water as he explored his new habitat. That first day, he stayed local, sticking to the coastal waters near the beach where he was released. On day two, he began venturing south, paying a brief visit to Cape Elizabeth, where he had first stranded back in January. By day three, it became clear that Dexxy's adventurous spirit had kicked in when he pinged twenty miles off the coast of Cape Cod. On day five, he was southeast of Nantucket. By day nine, Dexxy had already swum nearly one thousand miles from his release site to Georges Bank—a well-known marine habitat and commercial fishing grounds off the coast of Massachusetts.

As they collected the data, Marine Mammals of Maine shared a map of Dexxy's travels on social media. But after the day nine report, the updates stopped. A week later, I stopped by the rehab center to ask Lynda and Dominique what had happened. "It stopped transmitting," said Lynda with a look of disappointment.

There are a host of reasons why the tag might have stopped signaling: it could have fallen off or been damaged, its battery could have failed, or there could have been antenna issues that prevented it from connecting with a satellite. There was also a chance it stopped pinging because something had happened to Dexxy. I couldn't help wondering if the young seal, and the tracker, were currently in the belly of a great white shark. It was an unlikely scenario, however, as white sharks don't typically venture north from their winter habitat off Florida and the Carolinas until June, which was still a few months away.

"The tough thing is he was young and he headed to the continental shelf," said Lynda. "He was way the hell out there. He probably should have stayed a little closer to shore, but we can't ever know what happened."

For Dominique, there was another way to look at the limited data they'd received from the tracker. Only a very healthy seal could have covered that kind of mileage in nine days, she said, which meant the months of rehab support the team had provided had been effective. "There's only so much we can do on our end," said Dominique. "The rest is up to them."

A COUPLE OF YEARS ago, Kimberly was reviewing satellite data from the seal tags that had recently been deployed when she noticed something unusual. As she studied a map charting the locations of individual seals over time, she saw that one of the seals had been pinging from the exact same location for weeks. But the weird part was that the location was on land.

She entered the coordinates in Google Earth and zoomed in. She could see a house, and then a boat, realizing it likely belonged to a fisherman. She later discovered the house belonged to a commercial gillnetter who had mistakenly caught the seal in his net. Gillnet fishermen typically set their nets one day and haul them in the next. As fish are caught in the nets, they can attract the attention of seals, particularly gray seals, eager for an easy meal. But if the seal becomes entangled in the net while trying to consume the catch, unless the net is hauled up immediately, the seal will drown. Gray and harbor seals typically spend no more than thirty minutes underwater at a time before surfacing to breathe. While some seals are able to break free, they can sustain wounds or entanglements that will end up killing them later. While devastating for the seals, the situation is challenging for the fishermen as well. In the process of feeding on a fisherman's catch, seals can damage their equipment and gear. If a seal is accidentally caught and killed in a net, by law, the fisherman must throw the seal's carcass overboard. To Kimberly's relief, in this scenario, the fisherman had spotted the satellite tag on the seal and, presumably thinking the tag's owner might want it back, removed it from the seal and brought it home.

Kimberly reached out to Andrea "Dre" Bogomolni, a research scientist who was also based in Woods Hole, and chair of the Northwest Atlantic Seal Research Consortium. For years, Dre and her colleague, Owen Nichols, director of marine fisheries research at the Center for Coastal Studies (the same Owen Nichols who had encountered B-563, a gray seal that had been tagged on Sable Island, back in the late nineties), had been partnering with fishermen in Cape Cod to try to find safe and effective ways to deter seals from feeding on their catch. When Dre heard about the missing tracker, she immediately called one of her fisherman collaborators. No questions asked, she told him, but if he could find the fisherman in the fleet who had the tag, Kimberly would happily offer a modest reward to get it back. Twenty-four hours later, Kimberly had the tag in hand.

Fishermen and seal scientists have historically been at odds over their perspectives on seals, but that division is part of what Dre, Owen, and the scientists and fishermen they've teamed up with in Cape Cod are trying to address. Under the terms of the Marine Mammal Protection Act, fishermen are required to report every unintentional death or injury of a marine mammal caused by their operations, referred to as "bycatch." But the means to regulate this is limited at best, nor is there any incentive for fishermen to accurately report bycatch. "I'll just say it's pointless," Dre told me during a video call. "It's pointless. Because who's going to self-report? Nobody. And that's been proven over and over."

But getting an accurate understanding of how many seals are being killed in nets and other fishing gear is critical to understanding and protecting the health of the population. To address the issues around self-reporting, the National Marine Fisheries Service hires seasonal "observers" to board commercial fishing vessels and record what's caught and what's discarded, which includes bycatch.

Each year, scientists determine a "potential biological removal," or PBR, based on seal population estimates and observer data.

The PBR level is essentially a bycatch limit—the number of seals that can be unintentionally killed each year without risking the health of the overall population. "That calculation is used by the U.S. government as the tipping point to when management action needs to take place," Kimberly explained. When bycatch estimates near or surpass this figure, new restrictions are placed on fisheries to help protect the marine mammal population at risk.

Bycatch is a serious threat to gray seals. In fact, gray seal bycatch is higher than that of any other marine mammal species in the U.S., largely due to the New England gillnet fishery. "If a seal has an interaction with a gillnet, it's almost always not going to end well," Owen told me on a call. "When you think about animal welfare, it's a pretty awful way to go."

In 2017, an estimated 930 gray seals in the U.S. were killed through bycatch. In 2019, that figure shot up to more than 2,000 seals. It was the first year the number of seals killed as bycatch, as reported by NOAA, exceeded the PBR threshold.

When Dre first told me about the seal bycatch issue, I was surprised I hadn't already heard about it. It didn't seem to make sense that people were advocating for a seal cull if significant numbers of seals were already being killed through fishing operations. "It's an issue to the point where we should have management in place to reduce it," said Dre. "But we don't, because nobody talks about it."

Over the years, Dre, Owen, and others have submitted grant applications to NOAA Fisheries proposing to test methods to reduce bycatch by collaborating with local fishermen, but not a single application has been approved. "There's two thousand freaking gray seals dying in nets," said Dre. "I'm just like, how is this not a priority? Who's making these priorities? And who's making the decisions?"

Years earlier, as part of his graduate research, Owen was working with weir fishermen in Cape Cod who were also dealing with the challenge of seal predation. The fishermen suggested attaching "security cameras" in the weirs to monitor seal behavior. They added a sonar system as well, which allowed them to see in the dark. To their surprise, seals were feeding in the nets even more than they thought, and often at night. According to Owen, the issue wasn't just that the seals were eating fish and squid; it was that they were changing the behavior of the fish by causing them to rapidly change direction and leave the nets. They were disrupting the way the nets operate.

Owen worked with the weir fishermen to test a series of seal deterrents—many were similar to those trialed at the Ballard Locks. And, as was the case at the locks, the deterrents would work for a time, until the seals figured out a way around them. In the end, the only method that seemed to work was for the fishermen to haul up their traps multiple times per day, instead of just once, but that was enormously labor intensive and therefore unsustainable. They hauled their last trap in 2021.

Seal predation is unquestionably a challenge for fishermen, but it's certainly not the only threat they're facing. "It's death by a thousand cuts," said Owen. Ecological changes, such as warming ocean waters, can reduce the availability of fish. They're also contending with a rapidly shifting regulatory environment, the increasing cost of living, the price of diesel fuel, the inability to hire crew. And small-scale, community-based fishermen (as opposed to well-funded, lobbyist-backed industrial fishing fleets), have been the most severely impacted.

NOT LONG AFTER DEXXY disappeared off the grid, the remaining two gray seals at the center, Titan and Breaker, were due up for release. For Titan in particular, it had been a challenging road to recovery with a fractured flipper. Lynda decided on a private release, to give him plenty of time and space to transition from life in the rehab pool into the wild ocean without the added confusion of human spectators. But since she knew I'd felt a special connection to Titan, she invited me to join the team. I couldn't explain why I was partial to this particular seal. Perhaps because he was the mildest mannered of the feisty gray seals I'd met. Or perhaps because his big gleaming eyes seemed to stare right into me whenever I visited. Whatever the reason, I was rooting for him.

Because of Titan's injury, the two seals had never interacted before the day of their release. Instead of heading toward the water, as soon as they emerged from their respective kennels, they shuffled directly toward each other. For a few minutes, the seals sniffed and circled each other, seemingly fascinated by what they saw. It was as though they were looking into a mirror for the first time and dazzled by the beauty of their own reflection. Finally, the two seals turned and galumphed, side by side, into the sea. I silently warned them to steer clear of any fishing nets, and to keep an eye out for their pal Dexxy. I mean, you never know.

12

The Seal Snatcher

IN THE MONTHS I'D spent visiting the seal rehab center, I'd grown somewhat accustomed to the staff's Herculean efforts to balance seal care, research, outreach, hotline support, and stranding responses, not to mention the standard administrative and fundraising tasks inherent to any nonprofit. But nothing could prepare me for my first visit in the heart of harbor seal pup season.

By late May, months of dreary wet weather finally gave way to warm, sun-drenched days. Ospreys had returned to their nests, black flies had made their first unwelcome appearance, eastern phoebes were furiously delivering food to their chicks, and everything seemed to be in bloom. Meanwhile, the Marine Mammals of Maine team had doubled in size thanks to the arrival of five summer interns. On the morning of my visit, I expected the office to be buzzing with activity, but when I walked in, Fin and Stubby were the only ones there to greet me. I bent down to pat the dogs, then dropped my bag on the couch. I kicked off my sneakers, pulled on my rubber boots, slid a small notebook and pen into the back pocket of my jeans, and made my way to the animal care center.

When I opened the door, a cacophony of coos, bays, and howls echoed across the facility, emanating from a choir of at least a dozen harbor seal pups. The seals were stationed in groups of two or three in every available pool, tank, and fenced-in enclosure in the center. One by one, the pups popped their heads up from their respective abodes like cuckoo birds announcing the time: fish o'clock.

Marine Mammals of Maine is federally permitted to house up to sixteen seals at the rehab center at any given time. But with thousands of pups born each year along the busy shoreline, many end up needing urgent care, and Lynda and her team are forced to make tough decisions on which seals to admit. She makes a point to prioritize cases involving human interaction. Some of the harbor seal pups that wind up at the center are just a day or two old when they arrive and require feedings every few hours. This means a lot of late nights and early mornings for the rehab support team, all on a shoestring budget.

The staff and summer interns were scattered across the center, rapidly spinning from one task to the next—hosing down equipment and tanks, washing scat-smeared towels, prepping tubes and formulas for the upcoming feeding. They dodged and weaved between one another in an impressively choreographed caregiving routine. As I watched the scene, Lynda suddenly sped past me, her arms cradling a pup partially wrapped in a towel, like the Flash meets Florence Nightingale. "It's been a morning!" she called out over the whir of the industrial-grade dehumidifier. "Any chance you could help Katie record the intake?"

I looked up to see Katie already standing in the doorway to the triage room, clipboard in hand. The team was so flat-out with seal care that she had no choice but to wait for the first available human who could help. I quickly dipped my boots in the footbath to wash off any unwanted pathogens before entering the animal care space, and rushed toward the triage room. I pulled on a pair of latex gloves as Katie instructed me on how to prep a set of vials for bloodwork. She then dictated the newest patient's condition as

I carefully recorded the notes on the various admit forms. We were moving so quickly, I had barely registered the tiny harbor seal wrapped in a towel on the table in front of us. He had just arrived from Rockland.

As soon as we finished, Dominique appeared in the doorway. "I need to steal you," she said. "We just got a call on the hotline about a stranded seal pup on Bailey Island." A stranded seal wasn't exactly an appropriate thing to be excited about, but I was thrilled I might be able to witness a rescue firsthand.

Dominique finished up a few urgent seal care tasks as I ran back up to the office to grab my things, then headed to the front of the building to meet her. As I waited, Lynda joined me on the front steps to catch up briefly before we left for Bailey Island. "Are you hanging in okay?" I asked. "Honestly, Alix, I'm so tired," she said.

The month of May had been a chaotic one for Lynda, kicking off with a minke whale stranding in York, a coastal town in southern Maine. The whale had died on its own shortly after stranding, but the process of hauling the whale carcass off the beach (which required the volunteer support of Lynda's two nephews, who arrived with a flatbed truck and other equipment, courtesy of the Doughty family construction business), took hours. In Southern states, whale carcasses are often buried right on the beach, but Maine's rocky coastline prevents such an option. The massive carcasses must be transported to a nearby compost facility, where Lynda can conduct a necropsy to help determine the cause of death. By the time she got home, it was three in the morning. A few hours later, she was back at the rehab center to administer feedings for the youngest harbor seals.

A week after the whale stranding, Lynda journeyed even further south for a whirlwind research trip to assist Kimberly and several other seal research colleagues in their efforts to capture and tag gray seals on Nantucket. After a full day of fieldwork, during which the team successfully collected samples from eight juvenile gray seals, Lynda hopped aboard a ferry back to the mainland. She didn't get home until one a.m.

On top of this, the team had recently hit a rough patch—a large number of harbor seals had been arriving in dire condition. Just that morning, Lynda had euthanized one of the pups. As she told me about the case, her eyes filled with tears. "I thought we could turn things around, but she just wasn't improving." When Lexi Wright, the team's community engagement coordinator, had arrived at the office that morning, she found her boss sobbing at her desk.

"You'd think I'd be used to it by now," said Lynda, as we sat beside each other on the steps. Across the quiet road, a man unloaded a kayak from the roof of his car. Behind him, the Androscoggin River flowed by, oblivious to the dramas unfolding within the seal hospital.

Months back, I'd asked Lynda why they assigned numbers instead of names for the seals. Part of the reason, she explained at the time, was the sheer volume of strandings they respond to each year—four hundred, on average. But another reason was that many of the animals that are in rough enough shape to require rehab don't survive. I was lucky to have met several seals healthy enough to be released. Some, like Dexxy, had even received special names as part of their sponsorship program with local businesses. But as I'd discovered, these seals were the exceptions.

THE MIDDAY SUN WAS warm and bright as Dominique and I drove across the Cribstone Bridge to Bailey Island. Despite the focus of our mission, I was buzzing with excitement—not only was it my first ride in the Seal Mobile, but it was my first official stranding response.

Earlier that morning, a caller had reported that a harbor seal pup had been seen swimming and resting on ledges at the northern tip of the island for several days, with no signs of its mother. "I wish they'd called us sooner," said Dominique. "But they said the pup has been swimming, so hopefully that's a good sign."

As Dominique turned into a parking lot near to where the animal had last been seen, she suddenly went quiet.

"Oh my god," she said. "He's holding the seal."

I looked up to see a man who appeared to be in his midthirties, oddly dressed in a bright yellow wet suit, standing under a tree with a tiny seal clutched under one arm. His other arm was extended awkwardly. He was taking a selfie. The pup was squirming and crying out like a baying hound. A few onlookers stood nearby, seeming uncomfortable about the interaction but unsure of what to do.

As Dominique and I jumped out of the truck, the man in the yellow wet suit looked up, quickly dropping his selfie arm. "Remember me?" he asked. "I'm the guy who saved the seal last year." Dominique paused briefly as she absorbed the scene, then launched into a litany of questions to gather the case details: When was the seal first observed? What was the evidence of abandonment? What was its recent behavior? Simultaneously, she ushered the animal into a large kennel in the back of the truck. Its new name would be Number 80.

Once the seal was free from human arms, she turned to face the man and calmly but firmly reminded him that handling a seal, or any other marine mammal, is a federal offense. He rolled his eyes and shrugged. He then dramatically held his wrists together as if presenting them for handcuffs, and asked if she wanted him to put it back. "That's not helpful," she replied with remarkable restraint. "We're taking this seal. I'm just asking that next time this happens, you call us first. We're happy to come out and assess the animal."

"Fair enough," the man said. Bystanders thanked her for coming as we got back in the truck, the pup crying out from its kennel. My heart was racing as we drove back to the rehab center. I asked Dominique how often she's seen cases like this. She said they're far less common along the Midcoast than around Maine's more crowded southern beaches, but they do happen.

Last summer, for instance, a man paddling his kayak off Harpswell encountered a seal pup he believed nearby gulls were eyeing too closely. He lifted the seal into his kayak and brought it back to his house. He attempted to feed it, but the seal refused his offerings, which included cat food, milk, and a striped bass larger than the pup itself. By the following afternoon, when it still wouldn't eat, he called Marine Mammals of Maine to ask for advice on what to feed it. Lexi had been on hotline duty that day. ("He told me it was in his bathtub at the time," Lexi recalled when I asked her about it later, "so I asked him if there was water in the tub. He said, 'No, I'm not trying to baptize the thing.'")

Within hours, Lexi and another staffer had collected the emaciated pup from the kayaker's house and brought it back to the center, where it was immediately put on an IV to rehydrate and given time to rest. The team tended to the seal day and night, feeding it a specialized formula, antibiotics, and fluids as they attempted to counter its dehydration, malnutrition, and stress, but it was too far gone. The seal died after a week.

"It didn't have to die," Dominique said. She paused, and her eyes widened. "You know, I think it was the same guy. Didn't he tell us he 'rescued' a seal last year?"

As soon as we got back to the center, Dominique ran inside to ask the team to prepare the admit room for a new patient, and to debrief with Lynda about the case. I walked to the back of the truck to wait with Number 80. The tiny seal cried out helplessly from his kennel. I turned to look at him. From behind the cage door, his liquid eyes stared out at me and something lurched deep in my gut. It was a look of desperation, fear, pain. I knew I was projecting my own emotions onto this creature, and I could never know what he was feeling or experiencing. And yet, it seemed so clear in this moment.

I marveled at Lynda and her team of strong, resilient women—at their ability to manage this heartbreaking world of animal care. In these busy months, they spent nearly every waking moment struggling to keep creatures who don't speak our language alive, to find the right mix of food, rest, medication, and movement to revive them. And in the process, at least in the happier cases, they bear witness to that revival—their recovery from exhaustion, starvation, dehydration, as they gain strength and energy, while revealing their unique personalities. A seal brought back to life by the very same species that nearly destroyed it. It was evidence of our remarkable power as humans—to destroy and to save.

And yet it seemed to me that, in Lynda's case at least, it would be nearly impossible to manage such intensive animal care and simultaneously be so embedded in the research realm—a world where emotional detachment in the name of science was preferred, even required if you wished to be taken seriously within the scientific community. But Lynda's research colleagues, including Kimberly, told me that her contributions to their fieldwork efforts were invaluable. She had a unique ability to read the animals—to understand in moments of high stress what they needed, how to handle them, and when to back off.

Number 80 continued to wail. I felt helpless. "Don't worry, little guy," I said, in the same voice I use with my dog during high-anxiety vet visits. "You're gonna be okay." Of course, I had no idea if he would be okay, nor did I believe that my talking to him might assuage his fears in any way. But something compelled me to speak to him, as though the seal might hear something in the tone of my voice that would reassure him he was in the right place. Help was on the way.

For a brief moment, he stopped crying. He watched me, listened to my strange sounds. But the moment was fleeting, and he was soon crying again with renewed enthusiasm.

Shortly after, we hoisted the kennel out of the truck and carried him to the triage room, where Katie had already prepared an IV and a fresh set of intake forms. Dominique gently wrapped Number 80 in a towel and lifted him onto the examination table.

I looked over to see Lynda standing in the doorway. She was uncharacteristically silent, her eyes on Number 80 as she studied his condition. Katie handed her a stethoscope. After a brief examination, Number 80 was placed in one of the empty pools, and Lynda ushered us out of the room. He needed more fluids and formula, but more than anything, she said, with a look of concern on her face, he urgently needed rest.

ABOUT A WEEK LATER, I stopped by the rehab center to check in on Number 80. But I was too late. The seal pup had died that morning.

Given his condition, Lynda and Dominique told me that it was possible he had been abandoned by his mother before any human interacted with him, although handling a stressed and vulnerable seal while taking selfies certainly didn't help his survival odds. It's also, as Dominique noted, illegal.

According to NOAA Fisheries, violators of the Marine Mammal Protection Act can face over thirty thousand dollars in civil penalties, or up to a year in prison along with criminal fines, depending on the severity of the crimes. Lynda and Dominique reported the incident to NOAA's Office of Law Enforcement, just as they had done the prior year. While the individual was now a repeat offender when it came to seal harassment, his motives weren't nefarious—he wasn't actively trying to harm the seal, although he did seem to have a seal savior complex. The challenge for the Marine Mammals of Maine staff is that it can sometimes be the people who love and want to help seals the most who end up doing the most harm.

Between managing frustrated people who hate seals, or at least worry about their impact; people who love seals but are frustrated that Marine Mammals of Maine, which only has so much space in its rehab facility, can't help every seal in need; and the frustrations of the seals themselves—who in all likelihood have no idea what's going on but generally seem to be a feisty bunch—the challenges of the job at times can feel overwhelming, Lynda told me. But her efforts haven't gone unnoticed. In 2021, she was voted one of CNN's Heroes of the Year, after one of her volunteers nominated her. It was a moment that helped put Maine's seals, and the passionate, hardworking staff who care for them, on the map.

After Number 80 died, I asked Lynda how she stays motivated to do this work. If seal populations in the Northeast are thriving, and the outcomes of these cases are often tragic, why would she devote her life to saving individual animals? "Human interactions with these animals aren't going to go away," she said. "If anything, they're going to increase." And the cases where seals are being harassed—picked up by their flippers, thrown into the water when they're trying to rest—are the ones that seem to steel her resolve. Instead of feeling hopeless, she feels even more certainty that what she's doing to protect seals is necessary. "They didn't sign up for this—they just picked the wrong beach."

On the drive out to Bailey Island with Dominique earlier that day, I'd been reminded of the events that had inspired my own seal research. Just a mile from where we'd collected Number 80 was Mackerel Cove, where Julie Holowach had been killed by a great white shark three years earlier while swimming with her daughter.

I asked Lynda how she'd first learned about the attack. As it turned out, one of her friends, an EMT, had been one of the first responders at the scene. He called Lynda to warn her about what had happened, suspecting (correctly) that she was about to be inundated with frantic calls and media inquiries.

She was devastated by the news of Holowach's death, and heartbroken for her family. As she recounted the chaotic events of that day, she paused and shook her head. "I'm not looking forward to shark season."

As I drove home that afternoon, the signs of the changing seasons seemed to be everywhere—from the mild weather and longer days to the sudden increase in road traffic. And then, of course, there was Larry, the seventy-foot-long, seven-hundred-pound inflatable lobster that appears on the roof of the Taste of Maine restaurant near my house to mark the beginning of summer.

With Larry back in the neighborhood, it wouldn't be long before shark season was upon us. In June, great white sharks begin their annual migrations from their winter habitat off the coast of Florida and the Carolinas as they journey north to their feeding grounds. About two hundred miles south of Midcoast Maine, as the crow flies, was the largest aggregation of white sharks in the Atlantic Ocean.

I had one last research journey to make.

PART FOUR
SUMMER

And only then, when I have learned enough,
I will go to watch the animals, and let
something of their composure slowly glide
into my limbs; will see my own existence
deep in their eyes, which will hold me for a while
and let me go, serenely, without judgment.

—RAINER MARIA RILKE, "REQUIEM FOR A FRIEND,"
TRANSLATED BY STEPHEN MITCHELL

13

Pinnipeds as Prey

I'VE LONG BELIEVED THERE'S a magnetic force that draws humans to the ocean. And in mid-June, as I waited impatiently in stop-and-go traffic to cross the Sagamore Bridge to Cape Cod, I became even more convinced.

It's hard to imagine a time when summer trips to the beach weren't a societal norm. Yet beachgoing wasn't always an American pastime, and beaches were once considered places to fear, not to frequent. The coast was wild, unpredictable—a place of rogue waves, rough weather, shipwrecks, pirates, and unimaginable sea beasts.

But in the eighteenth century, the British elite began flocking to the sea in pursuit of the restorative healing powers of salt water and sea breezes. "Sea bathing" (which often referred to a rapid dunk in the ocean, as few knew how to swim) became a doctor-prescribed treatment for a wide range of ailments, including melancholy, tuberculosis, impotence, and "hysteria." Beachgoing soon gained popularity—not just in Europe, but around the world. In the late-nineteenth century, seaside tourism made its way to the Americas, with the first resorts emerging along the coast of New England.

By that time, we'd already doubled down on our efforts to

empty the oceans of the creatures that had once thrived there—whales, sharks, fish of all shapes and sizes. Seal bounties kicked off in Massachusetts and Maine in the same years that seaside tourism was first taking shape. By the mid-twentieth century, we'd systematically removed the wild from our coastal waters. In the process, we effectively created a giant saltwater swimming pool. There were risks, of course—riptides and jellyfish to be mindful of—but it wasn't long before swimming, surfing, and seaside recreation became a part of the cultural identity of America.

Today, more than 80 percent of the U.S. population lives in coastal states, and nearly 40 percent lives in the counties directly adjacent to the oceans and Great Lakes.

THE BUSY SUMMER TOURIST season had officially ended in Cape Cod, but the promise of clear skies, four-foot waves, and mild ocean temperatures still drew dozens of surfers to Newcomb Hollow Beach on a sunny Saturday in mid-September of 2018. Among them were friends and fellow boogie board enthusiasts Isaac Rocha, sixteen, and Arthur Medici, twenty-six. They had driven south from Revere, a town on the outskirts of Boston, the night before.

The sun was high overhead as Rocha caught a wave. After riding it to shore, he began paddling back out, diving under the surf as he swam toward Medici. But when he emerged, he heard a terrifying scream. Scanning the water, he spotted Medici—the sea around him stained red with blood. Rocha then saw the flash of a fin cut through the water, and he called out to Medici in a panic as he began rapidly swimming toward him.

Medici's head was face down in the water by the time Rocha reached him. He quickly positioned himself behind his friend, placed his hands under his arms, and began pulling him toward the beach. Surfers and swimmers—some of whom had witnessed the attack from the water—rushed to the shore, while lifeguards and medical staff sprinted across the sand toward Medici. One responder, noting the gaping wounds on his legs, applied a tourniquet; others performed CPR. When the ambulance arrived shortly after, responders still hadn't found a pulse. Medici was lifted onto a stretcher and hoisted into the back of the ambulance. During the thirty-minute drive to Cape Cod Hospital in Hyannis, paramedics continued their efforts to revive him, but it was too late. Medici was pronounced dead at the hospital.

It was the first fatal shark attack in Massachusetts in eighty-two years.

FOR YEARS LEADING UP to the fatality at Newcomb Hollow Beach, scientists had been studying the increasing presence of great white sharks, or white sharks, as they typically refer to them, aggregating along the Cape Cod coastline. White sharks are opportunistic eaters and consume a wide variety of prey, including fish, porpoises, and dolphins, but their increased presence in nearshore waters along the Cape's crowded beaches has been linked to the abundance of gray and harbor seals, particularly gray seals as the far meatier of the two species. "They're a predictable food source for sharks," said Megan Winton, staff scientist for the Atlantic White Shark Conservancy in Cape Cod. "We've seen them eat dogfish, fluke, other things, but they're not coming into the surf zone off the Cape to eat those things," Megan told me on a call. "The seals are the motivation there. If you think about it, they get more calorie bang for their buck if they take down a seal."

While white sharks have been notably absent from New England waters over the past fifty years or so, they are by no means new to the region. In 1849, Henry David Thoreau embarked on the first of three journeys he would make in his lifetime to Cape Cod, traveling southeast from his home in Concord, Massachusetts, to "get a better view than I had of yet of the ocean . . . of which a man who lives a few miles inland may never see any trace . . ." During his explorations of the sandy peninsula, Thoreau recounted his conversations with local fishermen: "They will tell you tough stories of sharks all over the Cape, which I do not presume to doubt utterly—how they will sometimes upset a boat, or tear it in pieces, to get at the man in it."

Evidence of white sharks in New England and Atlantic Canada dates even further back. Shark teeth over one thousand years old have been discovered in oyster shell middens.

In 1960, Harry Goodridge, of Andre-the-Seal fame, harpooned a nearly twelve-foot white shark after it consumed another of his "pet seals," Basil. In his book, *A Seal Called Andre*, he wrote that on August 12, he ventured out on his boat with the specific intent to hunt great white sharks, "with Basil along for company." But when they neared Mark Island, "I noticed that Basil's place in the bow was vacant. I wasn't unduly alarmed. He'd been restive and I figured he'd gone in for a quick dunk." Moments later, Goodridge saw blood in the water. He managed to harpoon the shark and after a two-hour battle, he wrote, he succeeded in exhausting the animal, killing it, and towing it back to the harbor. Photos of Goodridge standing beside the 1,200-pound shark were published in local newspapers. When he cut into the shark's stomach, he discovered Basil "in three pieces," alongside another harbor seal carcass.

By the 1960s, after many decades of seal bounty hunts in Massachusetts and Maine, there were few harbor seals remaining in New England waters, and gray seals had disappeared entirely. Without this predictable prey, it was initially unclear what was luring these top predators to Maine's coastal waters. In 1960 alone, there were eight white shark captures, including Goodridge's, reported near Mark Island—the largest number recorded for any year along the Atlantic coast.

After her husband's death, Goodridge's wife, Thalice, told a researcher that he had harpooned three other white sharks in subsequent years, and that he'd referred to Mark Island as "shark alley." According to a 1998 report in *Northeast Naturalist*, Thalice speculated that the sharks were attracted to the area by a poultry plant in nearby Belfast, which expelled chicken guts into the sea. The article's author, Paul Mollomo, found the theory quite plausible, not only because of the large number of sightings in the region during that time, but also because there wasn't a single sighting after the chicken plant closed in the mid-1960s.

SHARKS HAVE EXISTED FOR 400 million years, predating humans, dinosaurs, and even trees—but today, more than a third of sharks and rays are at risk of extinction, making them the second most threatened class of vertebrates in the world, after amphibians. Humans kill an estimated 100 million sharks around the world each year, largely driven by demand for their fins. Shark fins are the key ingredient in a traditional Chinese delicacy known as shark fin soup, which is popular in some parts of Asia.

Thoreau, for his part, vastly underestimated our ability to destroy these fearsome predators. "The ocean is a wilderness reaching round the globe, wilder than a Bengal jungle, and fuller of monsters, washing the very wharves of our cities and the gardens of our seaside residences," he wrote. "Serpents, bears, hyenas, tigers, rapidly vanish as civilization advances, but the most populous and civilized city cannot scare a shark far from its wharves." As it turned out, a single film provided enough inspiration for Americans to devastate shark populations. Directed by Steven Spielberg, *Jaws* was a 1975 box office sensation that vilified white sharks as ruthless, man-eating predators, igniting our primal fears. The film, based on a book written by Peter Benchley, not only showcased the power of a good story, it revealed the ways in which humans' ever-shifting perceptions of wildlife can impact their very survival.

The film inspired decades of shark trophy hunting. In the years after the movie was released, dozens of shark fishing clubs and tournaments sprang up in seaside locations along the U.S. East Coast, according to José Castro, a shark biologist with the National Marine Fisheries Service. In a 2017 paper in *Marine Fisheries Review*, Castro explained that the film had created such antipathy for sharks that some of these tournaments included prizes for the most sharks caught and the greatest number of pounds of shark landed.

"A collective testosterone rush certainly swept through the east coast of the U.S.," George Burgess, former director of the Florida Program for Shark Research, told BBC News in 2015. "Thousands of fishers set out to catch trophy sharks after seeing *Jaws*. It was good blue-collar fishing. You didn't have to have a fancy boat or gear—an average Joe could catch big fish—and there was no remorse, since there was this mindset that they were man-killers."

Benchley, the book's author, was devastated by the retaliatory slaughter of these animals. He devoted the remainder of his life to conserving sharks.

While the Marine Mammal Protection Act established blanket protections for marine mammals, including seals, in the 1970s, federal protections for white sharks weren't established until the late 1990s. Since then, as shark populations have begun to rebuild, they've been spotted in increasing numbers along the New England coastline—their return coinciding with the rise in seal populations.

Despite its negative portrayal of sharks, *Jaws* had another, more positive impact on young viewers. It inspired the careers of countless shark scientists and marine biologists worldwide. Greg Skomal, an internationally recognized white shark expert based in Cape Cod, is one of these scientists. Some have even likened Greg to a modern-day Matt Hooper, the quirky marine biologist played by Richard Dreyfuss in the film.

Greg, the senior fisheries scientist for the Massachusetts Division of Marine Fisheries, began studying white sharks in the early 1980s, but he didn't encounter one near Cape Cod until 2004. Five years later, he tagged his first white shark in Cape Cod waters. Over time, more and more sharks began appearing in the area.

Today, the waters surrounding Cape Cod are home to one of the largest seasonal gatherings of white sharks in the world, and the first-ever white shark hotspot in the North Atlantic.

TWO WEEKS AFTER ARTHUR Medici's death, Wellfleet town officials scheduled a public forum to discuss the shark attack that had shaken the community. Days before the event, anticipating a heavy turnout, officials shifted the venue from the town's senior center to the Wellfleet Elementary School, arranging as many chairs as they could fit in the school's gymnasium. Hundreds of Cape Cod residents filed in that evening, filling the rows of chairs, with some attendees standing at the back. News crews and other media formed a perimeter around the audience, their cameras pointed toward the front of the gymnasium, where eleven panelists—a mix of state and local government officials, scientists, and public safety experts—were seated in a row. Microphones had been arranged on the tables in front of each panelist, and another standing mic had been set up in the center of the audience.

At six p.m., Janet Reinhart, chairwoman of the Wellfleet Select Board, called the meeting to order, and the voices of the crowd quickly quieted to a murmur. "Our town is very distraught and sad about the loss of Arthur Medici, which has affected all of us deeply," she said. "Tonight, we are here to discuss what happened and what may be done to improve safety on Cape Cod beaches."

The scene was eerily reminiscent of the town hall scene in *Jaws*, in which police chief Martin Brody stands before a group of angry townspeople to discuss the shark attacks that had upended the fictional tourist town of Amity Island. But in the very real town of Wellfleet that evening, there was one notable difference. The biggest source of contention wasn't sharks.

After each of the panelists introduced themselves, Reinhart invited members of the audience to approach the microphone. Gail Ferguson, a Wellfleet resident who often swims at Newcomb, was one of the first audience members to speak. Over the past few years, Ferguson had been observing a growing colony of seals in Chatham. "The seals are attracting the sharks. They're bringing them into the swimming areas. And I imagine that this pattern has to be disrupted."

As the meeting continued, a resounding frustration amongst the speakers began to take shape. The issue was clear—there were too many seals. Calls to "close the restaurant" for white sharks on the Cape grew more pronounced, with several residents openly proposing a seal cull. Killing off seals, they said, would effectively eliminate the presence of white sharks on Cape Cod beaches. In response, David Pierce, director of the Massachusetts Division of Marine Fisheries, explained that it simply wasn't possible to remove seals given existing federal regulations. "There's nothing the Commonwealth of Massachusetts can do, except to continue to do the research and continue to advise the cities and towns as to how to try to improve communication and public safety," he said. "And the minute someone tries to change federal rules or law relative to seals, or white sharks for that matter, there are many, many advocates for 'don't make that change.' So it's likely never going to happen, despite the fact that there has been a death."

By this time, Gail Sluis, a resident of the nearby town of Brewster, had approached the microphone. Sluis had been on the beach the day that Medici was killed and had witnessed the rescue attempts firsthand. She commended the lifeguards and first responders who did everything they could to save his life. She then turned to face Pierce. "I understand what you're saying, sir, about the seals," she said. "But just because there are advocates doesn't mean we don't try. The seal population on the Cape is way out of control. They're eating all of our fish. Now they're eating our children." Sluis paused briefly as audience members began to clap in response. "This was a beautiful young man who lost his life because we've been sitting doing nothing," she said, her voice trembling. "That poor boy looked like a bomb went off on him. No sharks or seals are worth a young man's life. They're just not." The applause from the crowd echoed across the gymnasium.

Among the panelists that evening was Dre Bogomolni. Over the past several years, Dre had been speaking extensively to fishermen and other community members throughout Cape Cod who were frustrated by the rapidly expanding populations of seals. She was well aware of the tension that existed between the region's finfooted and flat-footed residents. But she wasn't prepared for the level of anger and confusion she witnessed that evening.

For Dre, the proposal to cull seals was unrealistic, but not just for legal reasons. Nearly 30,000 gray seals resided along the New England coastline, including the region in and around Cape Cod, but they weren't an isolated group. Those seals were part of the much larger population in the Northwest Atlantic—one that extended from Nantucket and Martha's Vineyard up into the Canadian Maritime, with the vast majority of gray seals breeding on Sable Island. There was a high likelihood that any efforts to remove gray seals from Cape Cod—whether by hunting or another intervention—would be offset by an influx from Canada, where there were hundreds of thousands of gray seals.

Dre also considered the research pointing to seals' positive ecological impacts, such as nutrient transfer. She was eager to share some of the scientific findings that could address public confusion and squash the misinformation that was circulating. She waited for the right moment to interject.

Cynthia Wigren, CEO and cofounder of the Atlantic White Shark Conservancy, and Dre's longtime friend and colleague, was seated beside her that evening. Cynthia had reached out to Dre before the event to invite her to join the panel, as she understood the importance of having a seal biologist present to answer questions from the audience. Cynthia was no stranger to shark-seal-fish-human controversies in Cape Cod. Having spent years advocating for white sharks, a species people feared, she understood how complex conservation issues could be.

She also understood how deeply the tragic death of a young man had rocked this Cape Cod community. Over the past decade, as seal populations had continued to expand in New England, and as the presence of white sharks along crowded Cape Cod beaches had increased, life as Cape Codders knew it had changed dramatically, and seemingly without much warning. Surfers and swimmers who grew up frequenting Cape Cod beaches now feared for their lives when they entered the water, and some fishermen had sacrificed hard-earned wages as seals damaged fishing gear and destroyed their catch.

But in that moment, Cynthia read Dre's body language—she saw her friend's frustration building as fear and confusion had taken hold in the crowd. She quickly grabbed a pen and wrote two words on a piece of paper, which she then tilted toward Dre: *Just listen.*

Dre glanced down at the paper and looked up at Cynthia. She smiled and nodded.

When Dre later recounted that moment, she said that it had fundamentally changed how she viewed the world. From then on, she began to define her role not just as a scientist, but as a community scientist. She credits Cynthia for helping her to understand the value of creating a space where people feel comfortable sharing. "I'm indebted to her," she said.

Dre was also grateful to Cynthia for understanding the value of having a seal expert on the panel, but she was frustrated that she had been the only person willing to serve in that role. "It was me when it should have been somebody federal," Dre later told me. "These are federally protected species, right? That's who should represent."

While there wasn't an "official" federal presence in the room that evening, Kimberly Murray was quietly seated in the audience. "I went sort of incognito," Kimberly told me on a video call. She listened carefully as one speaker after the next expressed their frustration with seals, but one of the most surprising moments for her occurred toward the end of the two-hour meeting.

Dana Franchitto, a sixty-seven-year-old surfer who lived in Wellfleet, had been surfing at Newcomb Hollow Beach the day Medici was killed. He told the crowd that he had also been affected by the growing numbers of white sharks in Cape Cod, and was now a much more cautious surfer. But he was morally opposed to the idea of a seal cull. "We're in their habitat," he said. "And I'm really appalled by this arrogant attitude that we have a right to play god in the ocean given our record on this planet." Some audience members loudly applauded as Franchitto returned to his seat.

"It was interesting to hear someone stand up and say that," said Kimberly. And it became clear in that moment that there were others in the room who felt similarly—a less vocal minority, perhaps, who were both saddened by Medici's death and also pleased to see the return of wildlife that had once thrived in Cape Cod waters.

In the wake of the fatal shark attack, a number of fishermen, surfers, and even local politicians, joined the call for a seal cull. One Cape Cod resident, surfer Karl Hoefer, formed an organization called the Atlantic Human Conservancy (a not-so-subtle nod to the Atlantic White Shark Conservancy) that aimed to convince Congressional leaders to overturn the provisions of the Marine Mammal Protection Act. In a letter to the editor of the *Cape Cod Times*, Hoefer wrote: ". . . Seals eat huge volumes of fish, putting fishermen's livelihoods at stake. My opinion is that seals should be hunted for bounty money, as they were up until 1972 . . ."

Emotions were running high. In a *New Yorker* article published just two days after the attack, Alec Wilkinson, who grew up spending his summers in Cape Cod and later worked as a police officer in Wellfleet, wrote, "In my childhood, I never saw seals, and it seemed desirable to protect them from being drowned in fishermen's nets. Now there are so many that one of my nieces described them as an infestation. This summer, I started to think of them as sea rats."

I'D ARRANGED TO SPEND a few days staying at a friend's uncle's house in East Falmouth on Cape Cod to explore the area, hopefully spot a few seals, and meet with some of the shark and seal scientists I'd been speaking to over the past year. The morning after my arrival, I drove east to Chatham to meet with Cynthia Wigren at the Atlantic White Shark Conservancy. It was a mild, hazy morning. Fog lingered on the treetops as I meandered along the back roads.

I met Cynthia at the Chatham Shark Center, one of two shark education buildings the Conservancy operates in Cape Cod. While the center was closed to the public that day, Cynthia met me at the entrance, greeting me with a warm smile as she unlocked the door and ushered me into the exhibit space. Tall with long, dark hair, Cynthia was serene, thoughtful, and soft-spoken. Her presence was an amusing contrast to the large-mouthed, aggressive-looking white sharks whose images and replicas filled the room.

We sat on a bench adjacent to a large metal cage, a version of those used for shark diving expeditions. An eighteen-foot white shark replica of "Curly," the largest recorded shark tagged off Cape Cod, hung from the ceiling above us.

Cynthia's shark career first began after a memorable experience cage diving with great white sharks in South Africa in 2010. When she returned to Massachusetts, she began to pay closer attention to the shark research occurring in her own backyard. She was particularly fascinated by the work of Greg Skomal in Cape Cod. When she discovered that Greg didn't receive any state funding to support his shark tagging research efforts, she googled "how to start a nonprofit." Not long after that, Cynthia founded the Atlantic White Shark Conservancy, where she currently serves as chief executive officer. Over the years, thanks to funding support from the Conservancy, Greg and his research team have tagged more than three hundred sharks.

Next door to the exhibit space, we could hear the collective laughter of the newly hired summer interns, who were midway through their orientation training. On the wall behind Cynthia, a model head of "Bruce," the shark from *Jaws*, appeared to bust out of the wall, his mouth wide and toothy.

"You can't talk about sharks without talking about seals," said Cynthia, as we discussed some of the community events that followed Medici's death. "I recognize there's so much negativity around seals. I think much more than sharks." While almost five years had passed since the attack in Wellfleet, Cynthia explained that within certain communities on the Cape, frustrations over rising seal numbers were still quite prevalent.

Earlier that morning, Cynthia, Greg, Megan Winton, and Marianne Long, the organization's education director, hosted a press event in which they shared some of their upcoming research plans. It was an effort to proactively engage with local media, to educate reporters about their shark research and help prevent misinformation from circulating. During the briefing, a local reporter had asked what was being done about the "exploding" seal populations on the Cape.

"The seals just seem to be the scapegoat," said Cynthia. "They call it an exploding population, but it's not. It's a recovering population coming back to its historic range."

For years, Greg, Megan, and their team had focused on the big picture when it came to studying white sharks in and around Cape Cod. Using satellite tagging technologies, they had spent nearly a decade monitoring sharks' movements and distribution, and more recently had begun to estimate their population size. But after the 2018 fatality, more and more questions were being asked about how sharks interact with their prey, specifically, seals. "In our minds, and in the minds of most researchers, white sharks don't bite people or surfboards or kayaks, normally," said Greg on a call. "It's a misinterpretation as a prey item, most likely a seal."

Shortly after Medici's death, the research team began deploying new technologies to better understand how sharks hunt—not just over the course of days to weeks, but over the course of minutes to hours—to understand their predatory behavior patterns. In 2021, the team determined that, based on their tracking data, white sharks off Cape Cod spend roughly half of their time in less than fifteen feet of water. According to Greg, this means that the animals are likely hunting seals for about half the time they're gathered along the coast. The findings offered critical information for beach managers to know sharks are present and actively hunting in many of the same areas where people are swimming and surfing.

"Sharks are very patient predators here," said Megan. "The seals know they're there, so they stay tight to the beach where they're just out of reach of the sharks. But the seals have to come out to eat at some point, and sometimes they get a little cavalier." With the help of drones, the team has captured footage of sharks slowly cruising along the outer sandbars near the beaches along the Cape as they wait for a seal to get "a little sloppy," at which point a shark will make its move.

"Any predator prey relationship is a very sophisticated game of cat and mouse," said Greg. "The seal does not want to be consumed, and it's not a stupid animal. Every move the shark makes, the seal's going to do a counter move. So it may be that the seal hugs the beach or stays on a sandbar, and what's the shark going to do in response to that?"

Recently, the team has begun to deploy a new tracking technology—camera tags. These small electronic devices are implanted just behind the shark's dorsal fin, providing scientists with a "shark's-eye view." The trouble is, the tags only stay on for a day or so before popping off the animal. And for the research team, it's impossible to control the timing of when the sharks are tagged and when they successfully take down a seal. Seals are by no means easy prey, even for highly skilled predators. Not only are they incredibly fast and agile swimmers, they're also large animals that aren't afraid to use their sharp claws and teeth as defense weapons. "A lot of the sharks that we get underwater footage of have evidence of seals fighting back," said Megan. "They have seal scratches around the eyes and sometimes down the body."

While the team has recorded several seal predation attempts, they're still looking for the smoking gun—footage of a successful hunt. In a few cases, they've gotten frustratingly close. A couple of years ago, Megan collected a camera tag that had popped off a shark they'd named Terp, but after reviewing the footage, she didn't find any evidence of seal predation. The very next day, however, while back out on the water tagging sharks, they happened across Terp once again. The shark was midway through a seal supper.

In 2019, the Cape Cod National Seashore and other local government groups commissioned a study to evaluate strategies aiming to increase public safety on beaches. The independent research firm they hired analyzed dozens of methods that had been proposed to mitigate white shark interactions—from seal culls and seal contraception, to shark alert technologies such as tagging, drones, and sonar detection devices, to barrier systems like exclusion nets that would keep sharks away from beaches.

The final report concluded that a silver bullet solution to address the white shark issue simply didn't exist. Even the most promising strategies were expected to have a limited impact at best. There was, however, one strategy that proved to be consistently effective at reducing white shark encounters, although it wasn't a particularly favorable one: changing human behavior.

After the report was released, the Massachusetts state government increased funding for Cape Cod communities to scale up their public safety programs as they prepared for potential shark attacks. Beachgoers signed up for free "Stop the Bleed" trainings, complete with a take-home tourniquet they could tuck into their beach totes.

AS LUCK WOULD HAVE it, my visit to Cape Cod coincided with the world premiere of a new HBO documentary, *After the Bite*, which focused on the impacts of Arthur Medici's death on Cape Cod residents. The film was being screened as part of the Provincetown Film Festival, and was slated to include a question-and-answer session with the director, Ivy Meeropol. I managed to snag a ticket.

The theater was packed; hundreds of people had shown up to see the premiere. The film itself was, in my view, fair and informative—a far cry from the sensationalist portrayals of shark attacks splashed across the Discovery Channel's Shark Week. But the most fascinating part of the event occurred after the credits rolled. During the Q&A, a young man seated toward the front of the theater raised his hand. He commended Meeropol for presenting such diverse viewpoints about the issues while also showing empathy to everyone—not just the humans who were profiled, but also the wildlife at the center of the conflict. His question was, how did she do it? How did she create a film about such an emotional and dramatic conflict without a single villain?

Meeropol was thoughtful in her response. She explained that she had recognized her power in the edit room to shape someone's story to fit a particular narrative, and she felt the weight of that responsibility. She'd wanted to present each person in the context of their own stories. She wasn't trying to make sense of the tragedy or to find some meaning or key takeaway to share with audiences. She just wanted to give different sides a voice—to *just listen*. There was no right answer because, for these community members, the experience was itself so personal. "It's not just an idea they're trying to convey," she said. "It's their lives. It's who they are."

Dusk was settling as I drove out of Provincetown that evening. As I turned onto Route 6, I stopped short to allow two wild turkeys, neither of which expressed any sense of urgency, to cross the road in front of me. A few minutes later, I hit my brakes once again, this time for a jaywalking coyote. It was still light out, and with no other cars around me, I wondered if I'd simply imagined these animals—a mirage of meandering wildlife.

Coyotes, wild turkeys, sharks, seals—animals capable of adapting to a world dominated by humans, characters in a story that was still being written.

14

Wild Cape Cod

IF CAPE COD IS shaped like a flexed arm in the sea, with Chatham at the elbow, then Monomoy National Wildlife Refuge would be the Cape's dislodged funny bone—a series of barrier beaches and islands extending eight miles off the coast. The refuge is a popular haul out site for gray and harbor seals in Cape Cod, so I decided to check it out.

I parked in a small lot beside a closed visitor's center. After spending several minutes searching for trail access, I nearly gave up. Every entrance I found had been roped off due to bluff and beach erosion. But then I noticed a small hut at one end of the lot with a sign that read, "Friends of Monomoy National Wildlife Refuge." Beneath the sign, a gray-haired man sat behind a table piled with maps and brochures. I asked him about the trails, and he explained the entrance had been rerouted. He pointed toward a paved road that was barely visible behind an intimidating collection of "Private Property" signs. "Ignore the signs," he said, knowingly. The refuge maintained a deeded right of way that allowed for public access, but that didn't mean the oceanfront homeowners were thrilled about the visitor foot traffic.

"Any chance I'll see seals down there today?" I asked. He shook his head and explained that they were mostly in the water looking for fish this time of year. "But I'll tell you a funny story," he added. About five or six years ago, he and a friend were fishing for striped bass at Harding Beach, not far from where I'd be walking that day. "I got a hit," he said. "The fish didn't take, but the seal saw me get the hit." The seal followed the two men as they walked back and forth down the beach, swimming parallel to them in the water, hoping for an easy meal. Finally, the men came up with a plan. They would walk in opposite directions, forcing the seal to choose who to follow. That way, at least one of them would have a chance to fish in peace. "But you know what that seal did?" he asked. I shook my head. "He stayed right in the middle." I looked at him, confused. "Because he couldn't make a decision?" "Oh no, he could make a decision. He decided to watch both of us." I laughed. "They're not stupid, those seals."

Despite the warmth and humidity that day, the beaches that extend along Monomoy's Morris Island trail were remarkably vacant. I was amazed by how different the landscape was compared to the rocky coast of Maine. Here, endless stretches of windswept dunes, white sand beaches, salt marshes, and freshwater ponds are surrounded by clear, turquoise-tipped waters.

After walking a short, scenic loop, I ventured off-trail down the beach, hoping to encounter at least one seal. About half a mile down, the beach curved sharply to the right, the path ahead obscured by the dunes.

As I continued around the bend toward the tip of the peninsula, I suddenly came across more than half a dozen fishermen—surf casters spread wide across the flats. The men were dressed in chest waders and long-sleeved shirts, with baseball hats shielding their faces from the sun. Their bait boxes were strapped to their waists, as they swung their lines overhead in tandem, flicking them forward into the surf.

I stopped walking to watch the nearest fisherman—from what I hoped was a respectable, un-creepy distance. He cast his line, and the bait landed with a soft plunk in the water. Seconds later, I gasped in surprise when a large, round shape emerged next to the exact spot where the bait had landed. The massive noggin and long, prominent nose were unmistakable—a male gray seal. The fisherman, for his part, didn't so much as flinch. He'd clearly been well aware of the seal's presence. The creature turned its head to the side as it scanned the flats, revealing the full extent of his horse-like honker. I was riveted, and began inching toward the edge of the water, watching the seal as the seal watched the fisherman.

Unfortunately, my curiosity caused me to breech the respectful line of separation between myself and the fisherman, and I was now firmly standing in a creepy-close observation zone. The fisherman reeled in his line, then turned to face me.

I quickly apologized for my intrusion, explaining that I was researching the population recovery of seals. "I couldn't help but notice you have a seal hot on your trail here," I said. The fisherman, who appeared to be in his sixties, rolled his eyes. "It's the cost of business, isn't it," he said, with an English accent. "But it's frustrating, there's just so many of them." He explained that he'd been fishing at Monomoy for at least twenty years—striped bass, mostly, occasionally bluefish—but it almost wasn't worth it these days with so many seals around.

He was also well aware of the presence of white sharks. Just yesterday, his friend had encountered a dead seal on a nearby beach with huge bite marks, presumably from a great white. But he didn't believe that sharks were making any dent in the seal population.

As we chatted, the seal began making the rounds to each of the anglers in turn, popping his head up each time a baited hook plunked. He moved with impressive speed, then disappeared beneath the water.

Suddenly, a fisherman standing further out on a sandbar began shouting. He'd hooked a fish, and was struggling to reel it in as quickly as he could. Just then, on the opposite side of the sandbar, two large eyeballs and a shiny round head popped up from the water like a periscope. The seal eyed the fisherman briefly, then quickly disappeared beneath the surf.

I imagined it jetting like a torpedo through the water, boomeranging around the tip of the sandbar in his attempt to reach the fish before it was reeled in. As I watched the scene, riveted, the fisherman I'd been speaking to turned to me. "Just wait for the explosion if the seal gets it," he said. At first, I assumed he meant an explosion of splashing water if the seal nabbed the fish, then realized he was referring to an explosion of fisherman fury.

The seal was seconds too late. The fisherman successfully reeled in a sizable striped bass. After measuring the fish with a tape measure, he tucked it safely in his cooler. He then stood and shouted at the seal, his words amplified by the wind across the flats. "Missed that one, didn't you, you fat fuck!" The seal stared back, silent.

Minutes later, I heard more shouting as another fisherman hooked a fish. This time, however, the seal was within easy striking distance. The fisherman shouted at him, emitting a series of yips and yaws as if herding cattle. But the seal's jaws had already locked onto the fish. He splashed briefly at the surface as he deftly ripped it from the line and disappeared underwater with his reward. The slackened line floated in the current.

As I walked back down the beach, pulling my hat lower on my face to shield it from the strong June sun, I considered the plight of the anglers. One seal competing with seven fishermen, nabbing fish right off their hooks. It was hard not to be impressed by the seal's performance, but I could certainly understand their frustration. If the seals had been here twenty years ago, these men probably wouldn't be here now. But they'd been coming to this beach for decades, spending quiet, meditative hours standing in the surf, reeling in dinner, doing what they loved.

Still, it was hard to blame a seal for being exactly what it is.

I also considered the plight of the fish. The population of striped bass had tanked in recent years, and quotas for fishermen had been significantly reduced. And yet, if you added up what the fishermen took home along with what this one seal pulled off their lines, the total "catch" would look quite a bit different. I'd heard similar concerns raised during my visit to the West Coast. Did the catch limits officials set for both commercial and recreational fisheries factor in seal predation that occurred as part of the human fishing exploits? In other words, were we underestimating the "take" and thus further impeding already vulnerable fish populations?

ON THE MORNING AFTER my Monomoy visit, I met Dre Bogomolni at a café in Falmouth. After a year of emails and multiple video calls, in which I'd harassed her with as many seal-related questions as I could think up, I was thrilled to finally meet her in person. Dre was in her midforties with thick, wavy brown hair and a warm smile. As chair of the Northwest Atlantic Seal Research Consortium, among several other research and teaching roles, she had a thorough and nuanced understanding of the benefits and challenges related to seals' recovery, particularly in Cape Cod.

One of the biggest sources of frustration for Dre over the years, when it comes to communicating information about marine species recovery, has been the ways in which the media present a "he said, she said" narrative, pitting scientists against fishermen. Shortly after the 2018 shark attack, Dre was frequently asked to speak to reporters about the increasing numbers of seals in Cape Cod. But she grew tired of the consistent misrepresentation of the issues.

"Before the interview, I'd say, 'So who's the fisherman?' They'd tell me. And then I'd say, 'You do know that we're working together, right?'" She smiled and shook her head. "I wish 2018 Dre was 2023 Dre, where I just will not budge now. And I don't mind calling people out, as long as I call them back in and give them a resource, right?"

IN THE SUMMER OF 2021, after Julie Holowach was killed by a white shark off the coast of Maine's Bailey Island, a similar media frenzy unfolded, with yet another onslaught of misinformation about seals. During that time, fisheries groups began widely circulating an Associated Press article that quoted a Maine-based shark scientist and included what were, in Dre's view, "outlandish" statements the scientist made about seals. The quotes weren't just misleading, said Dre, they were inaccurate. She'd hit her breaking point.

"I called the editors and I said, 'This is atrocious. It's really poorly done. I'm an expert in this field, and if you want to talk to someone about seals, I have a ton of resources, not just me. You people give me a call.'" She left a similar message with Reuters. Within minutes, one of the reporters called Dre, and the outlets were quick to retract and reissue the articles, using quotes and information sourced from seal biologists. "I made these beautiful connections with great journalists. They would then call me and say, 'I have a question, what should I do?' It changed the landscape of what was coming out—not just the story, but the headlines."

That same summer of 2021, a team of scientists led by Jennifer Jackman, professor of political science at Salem State University, conducted a study to better understand local perceptions of rebounding populations of seals and sharks. The aim was to understand how people perceived marine wildlife in Cape Cod, and how those perceptions may impact their relationship with marine species and ecosystems.

The researchers surveyed two thousand individuals from three groups of stakeholders on the Cape—residents, tourists, and commercial fishermen. The findings indicated that all three groups believed that sharks are important to the marine ecosystem. When it came to seals, however, the groups' attitudes differed. Commercial fishermen tended to blame seals for reducing and suppressing fish populations, hurting the economy, and creating public safety risks by attracting sharks. But most residents and tourists viewed seals favorably and hoped to see them in Cape Cod. Dre was happily surprised by these findings. "I thought, really? People like them?"

At the same time, she acknowledges that the frustrations reported in the media are very real, even if they're a minority. "The fishing community part I see," she said. "I see that every day. It's real. There's a spectrum of opinions and associations, depending on your experiences. And I get that."

In the end, Dre believes seals are too often being used as scapegoats for other issues, such as the regulatory challenges and frustrations plaguing fishermen. "It's a human issue," she said. "It's not a human-wildlife issue. It's a human-human issue. The wildlife is fine without us. That's my take."

The stakeholder research study revealed that roughly half of the respondents had no knowledge of the history of seals and sharks in the region, such as the impact, or even existence, of seal bounty hunts.

It's understandably easy to perceive seals as invaders if we have no memory of their prior existence, nor our role in their annihilation. Daniel Pauly, arguably the world's most influential fisheries biologist, put it best: "We transform the world, but we don't remember it. We adjust our baseline to the new level, and we don't recall what was there."

In the mid-1990s, Pauly coined the term "shifting baseline syndrome" to describe the ways in which we ignore the scale of environmental destruction by continuously redefining the reference point for a given species. In the case of Atlantic cod, for example, if fisheries biologists define, as a baseline, the population and size of the cod that existed when they first began their careers, then any change in that population will be measured against their early-career reference point. But what if these biologists began their careers in the late 1980s? The "baseline" population of cod in that case, after industrial fishing fleets had already demolished the stocks, would be a miniscule fraction of its historic size. The cod population today may have increased when compared to the 1980s stocks, but it's still a dramatic depletion from the far larger population that existed throughout history.

The same theory holds true when it comes to seals. Most adults today who grew up in New England did so in the near-total absence of these creatures. They perceive a balanced ecosystem as one in which seals don't exist. It's the same for white sharks, and for recovering wildlife across the globe. How can we hope to coexist with rebounding species if we have no knowledge of their historic presence, nor their ecological roles?

And yet, with new generations come new perspectives.

BEFORE MY CAPE COD journey drew to a close, I signed up for a guided seal walking tour offered by the Center for Coastal Studies, a nonprofit research group based in Provincetown. I was instructed to meet the group at Head of the Meadow beach in North Truro, part of the Cape Cod National Seashore.

Our tour guide was Jesse Mechling, the director of education at the Center for Coastal Studies. Aside from Jesse, we were a small group of three—myself and a friendly middle-aged couple from Maryland who were seasonal visitors to the Cape. They loved seals and were excited to learn more about their ecology.

Jesse led us on a ten-or-so-minute walk down the beach, until we neared a sandbar that jutted out into the surf. From there, the beach curved sharply to the left. He explained it was a popular haul-out site for gray seals. Unfortunately for us, the sandbar was vacant that afternoon, although we did see two shiny seal noggins bobbing in the water in the distance, looking toward the shore.

Jesse set up his spotting scope on the sand so we could get a better view of the distant seal pair. But as he adjusted the scope, the seal heads began to multiply. Soon, I counted more than thirty seals gathered side by side, bottling in the water, as they slowly drifted in our direction. It was as though they were midway through a committee meeting. Oddly, all thirty seals seemed to be looking at the same fixed point on shore, out of our view beyond the sandbar. I recognized that look.

A moment later, a man appeared at the bend in the beach, walking (though perhaps stomping would be a more accurate description) in our direction. As soon as we spotted him, the reason for the seals' intense focus was immediately clear—the man was holding a fishing rod. As he continued down the beach, the seals swam alongside him in a comically coordinated clump, thirty sets of wide ocean eyes glued on that rod. The man, who barely looked up as he marched past us, was clearly displeased with the situation and seemed to have abandoned his fishing attempts. The seals, taking the hint, eventually dispersed, although I wouldn't have been surprised if they assigned an underwater scout in case the fisherman dared to begin casting again.

For us seal tourists, however, it was an ideal scenario, as dozens of gray seals were now positioned right in front of us in the water. We took turns peering through Jesse's spotting scope, which offered such an extreme close-up, you could see inside each of their moth-shaped nostrils.

As we watched the seals, Jesse spoke about some of the controversies related to rising seal numbers. But he emphasized that most people on the Cape don't hate seals, citing the recent Cape Cod study findings. His view was that those with unfavorable opinions of seals tend to be far more vocal than their counterparts. He mentioned a guy in his neighborhood who drives around with a "Kill the Seals" bumper sticker. It's not that the anger and hatred isn't there, said Jesse, but when it came down to it, he didn't believe there would be enough support to instate any sort of seal cull. By way of example, he referenced an incident that had occurred thirteen years earlier.

IN THE FALL OF 2010, a severely injured gray seal was discovered stranded on a beach in Truro, not far from where we now stood. A team of marine mammal experts was called in to examine the animal. They determined its head wounds had been caused by gunshots. The seal, which appeared to be in considerable pain and had little chance of survival, was euthanized.

Months later, six gray seal carcasses were discovered on beaches across Cape Cod, all with evidence of gunshot wounds in their heads. Federal fisheries authorities assigned a special agent to investigate the shootings. The agent encouraged residents to share information that might be relevant to the case, while animal welfare groups offered a reward of fifteen thousand dollars for any leads. No one came forward. Meanwhile, the seal skulls were packed in boxes marked "evidence" and transported to a laboratory at the Woods Hole Oceanographic Institution. There, researchers conducted CT scans and necropsies to gather as much information as they could to inform the investigation. It was unknown how much time had elapsed between the shootings and the dead seals washing up on shore, and the data were limited at best. In the end, with little evidence beyond bullet fragments to work with, the agent was unable to identify a culprit.

But the news had prompted considerable public outcry in Cape Cod and beyond.

Growing up, Jesse never saw seals in Cape Cod, and everything he knew about white sharks he'd learned from the movie *Jaws*. He still remembers the very first seals he saw in the nineties, how exciting and rare that was. But Jesse's son has had a very different experience, he told us. His son had grown up with healthy populations of seals and sharks in the water. That was his baseline. If those animals were removed, he would see it as a loss, as an imbalance. It was all about perspective.

Conservation biologist Joe Roman refers to this concept as "lifting baselines." For generations, scientists and wildlife managers have been trained to manage species in decline. But now, for the first time in many of their careers, they're also contending with a number of recovering species, including seals. In 2015, Joe and his colleagues published an article in *Trends in Ecology and Evolution* in which they proposed the concept of lifting baselines as a subset of the shifting baseline syndrome, which tends to focus on downward trends. Lifting baselines is a way to describe the success stories related to species recoveries, as well as the associated conflicts and challenges.

In Joe's view, one of the biggest challenges when it comes to the return of wildlife like seals and sharks is that we didn't help coastal communities to prepare for it. New Englanders adjusted to life in the absence of these marine species. We grew accustomed to swimming in the ocean without fear of sharp-toothed predators lurking beneath the surface. We perfected our fishing casts along the shore, reeling in stripers and bluefish without the watchful eyes of blubbery, opportunistic competitors. And now that they've returned, now that life as we've known it has fundamentally changed, there's limited guidance for communities seeking to coexist with these creatures.

When the Marine Mammal Protection Act was passed, there wasn't a strong understanding of how quickly, or even whether, individual species might recover. But now that some populations have rebounded, coastal communities are grappling with how to address growing conflicts.

As Jesse pointed out, as new generations are born, perspectives are already shifting. Joe spoke to this as well. "The older generations who aren't used to seals and sharks, they're going to move or they're going to pass on," he said. "The younger generations are going to be used to living with them." Having thriving populations of seals and sharks will be part of the identity of the Gulf of Maine, of Massachusetts Bay. "My students today have no living memory of a time when these populations were close to extinction."

While driving back to Falmouth after Jesse's seal tour, I passed a sign on the road, just south of Truro, for Newcomb Hollow Beach—the same beach where Arthur Medici had been killed. On a whim, I took a left off Route 6 and headed for the shore.

It was nearly five o'clock when I pulled into the parking lot, which was surprisingly full. As I walked toward the path to the beach, I passed several prominent signs with information and warnings about great white sharks. A "Stop the Bleed" kit had been placed at the entrance.

I walked along the beach a short distance, then plopped down in the sand to stare out at the smooth surface of the water. Beside me, a young couple and their leashed dog paused to greet friends who were seated in a cluster of beach chairs and coolers. They then continued down to the water, where the packed sand made for an easier stroll. As they passed, I overheard the man, in a quiet, calm voice, say to his lady friend, "Oh, there's a seal."

I followed his gaze. Sure enough, a gray seal had popped his head out of the water, just a few yards from where the couple and their land dog now stood. The seal stared at them from the edge of the surf. The woman smiled and nodded, and they continued walking down the beach.

15

War and Peacebuilding

IN EARLY AUGUST, DOMINIQUE Walk lifted Number 39 onto a towel at the Marine Mammals of Maine rehab center, gently restraining the harbor seal as Katie Gilbert clamped a blue plastic tag onto the webbing of his hind flipper. The tag would ensure that the seal could be identified in the future, should a researcher, rescue group, or anyone else encounter him. After nearly four months of care and many happily gulped fish meals, Number 39 and his pool-mate, another young harbor seal, were ready for release.

Finding a suitable location for the release, away from roaming humans and dogs, was tricky during peak tourist season, but friends of Lynda's family had offered their private beach for the occasion. The staff carried the two kennels down a short path and placed them on the sand facing the water. As they unlocked the cage doors, Number 39's pool-mate immediately poked her head out to look around, then quickly caterpillar-crawled to the water and submerged herself. Number 39, meanwhile, balked at exiting his kennel. But a few minutes later, after the standard coaxing from Lynda (another gentle crate tip), he reluctantly emerged and slowly galumphed his way to the water.

As he cautiously examined his new habitat, I sat beside Lynda's mom, Carol, on a nearby rock to quietly observe the scene. We watched the seal playfully explore a tidepool, slapping the water with his front flipper and burying his face in rockweed. I asked Carol what it was like to be able to observe her daughter's work in this way. "I learn from her," she said, smiling. "Her father and I still don't understand how she did this."

The staffers were laughing and chatting with one another as the seals they'd spent months caring for returned to the wild. With the constant heartbreak, the exhausting hours, and the divided public opinion about their work, a pair of tallies in the win column clearly meant so much to Lynda and her team.

For a moment, the young seal looked up at the small crowd of humans celebrating on the beach. Then, turning his large, dark eyes to the sea, this curious, charismatic, wild-once-more creature slid out of the tidepool and into the surf, disappearing without a splash.

That evening, I considered Number 39's storied existence—still perplexed by the notion that the perception of a seal could differ so dramatically in the minds of the humans who interact with it. I was sitting on my friend's porch at the time, watching the late-summer sun filter through the trees across the yard, spinning green grass to gold. As we waited for the mosquitoes to descend and drive us inside, I shared with him the story of Number 39's early human encounters.

The pup had been born on a ledge along a craggy stretch of shoreline in Harpswell. His mother had selected the ledge wisely— above the high tide line, in a protected cove with few predators to worry about. She'd be able to leave her pup there for hours while she foraged and return later to nurse him. It was a perfect location, save for one detail—the human owner of the ledge wanted the seal gone.

The landowner, whose house was just uphill, was eager to have his dock installed for the upcoming summer season. The seal pup was in the way. The man was advised to call the reporting hotline for Marine Mammals of Maine. Over the next two days, staffers encouraged the landowner to let the seal stay, explaining that relocating him would prevent the mother from being able to find him, which would mean an almost certain death for the young seal, not to mention the stress on his mother when she returned to find her pup gone. But the landowner was undeterred, refusing to delay his dock installation another day. He insinuated that if nobody else showed up, he'd move the seal himself. Fearing for the safety of the animal, staffers collected the pup—the thirty-ninth stranding they'd responded to that year. He'd been the first harbor seal of the pupping season admitted to the rehab center.

As I recounted the details, I felt myself becoming indignant. I didn't understand how someone could have knowingly put an animal's life at risk over what seemed to be such a minor inconvenience. Marine Mammals of Maine needed to raise upward of ten thousand dollars to cover five months of care for the newborn harbor seal. And then there was the moral argument, and of course the complex logistics, the late-night and early-morning feedings for the already stretched staff. And then there were the other seals to consider, the ones that couldn't be admitted because there wasn't a space available.

Midway through my rant, my friend suddenly jumped out of his chair and ran to the edge of the porch, clapping and shouting. Startled, I turned around to see what was happening. About ten yards from the house, a large deer was pulling leaves off his newly planted honey crisp apple tree. The deer, caught red-hoofed, looked up, nonplussed by the noise, then immediately resumed his munching.

"Why is he doing this?" he asked, incredulous. "Why can't he wait until the fruit comes? The deer are killing the tree that could feed them *later*." It took everything in me to maintain a straight face. "I don't think deer spend much time planning for the future when it comes to eating," I said. "They're more of a 'live in the moment' kind of creature." "Then they're being dense," he said, with a half smile.

There were wild apple trees all over the yard that the deer hadn't touched, but as soon as the young, cultivated apple tree had been planted, the deer immediately stripped all the leaves off the branches, then ate the branches themselves. They'd damaged the tree to the point where it was unlikely to bear fruit for years, and would be lucky to survive the winter.

For my friend, this felt like a personal affront. He was fine with the idea of sharing his apples with the deer, but he was infuriated by the notion that they were destroying the tree itself.

Yet there was something so incredibly *human* about their behavior—the immediate payoff, the unsustainable consumption habits. As I casually pointed out the irony, he reminded me that it wasn't too long ago that I'd threatened to shoot any deer that dared interfere with my vegetable garden. In my defense, I'd made that comment after spending five hours hauling wheelbarrows of soil to build the garden beds, and—as it was my very *first* vegetable garden—I was feeling new-plant-mom levels of protection over my seedlings. (Beyond that, I didn't own a gun, and promise I had no real intention of shooting backyard wildlife.) But I took his point. I would have been furious if the deer had destroyed my garden. I'd purchased soil, compost, stakes, wood, and seedlings, not to mention the sweat equity I'd invested.

Neither of our situations presented issues related to food security. I could buy fresh vegetables at a grocery store or farm stand, and he could drive five miles down the road to an orchard that sells the best honey crisp apples in all of Maine. We both knew, rationally, that the deer were simply being deer.

And yet, on some level, the deer's behavior felt disrespectful.

The animal had trespassed and destroyed personal property. Worse, it had stared right at us while doing it, as though it felt no sense of guilt or remorse over its wrongdoing. Imagine.

A FEW DAYS LATER, Lynda sent me a message to let me know that Katie would be performing a seal necropsy, the equivalent of an animal autopsy, within the hour. If I could get there in time, she said, Katie could use some help with the notes. The staff remained flat-out managing a full house of demanding seal patients.

I'd been asking Lynda for months if I could observe a seal necropsy, but my request was usually met with a polite smile and the question, "Are you a fainter?" But I was eager to understand how scientists collected data to determine the cause of a seal's death, how that information could inform future rehab efforts or improve our knowledge of the health of the population at large.

I responded with enthusiasm, hoping the alarming number of exclamation marks in my reply didn't cause her to question her decision. Twenty minutes later, I pulled into the parking lot at the center. Katie was already outside, hunched over a long, stainless-steel table with a large basin hanging down from one end. She was dressed in what looked to be a cross between a surgical gown and a rain poncho, along with a mask, face shield, and latex gloves.

As I approached, she looked up from behind her face shield. She then stood and lifted her hands vertically in the air, her gloves covered in seal blood. "Hi Alix!" she said, cheerfully.

The seal carcass was splayed across the table, with several of its organs already dissected and arranged in neat rows, prepped for analysis. The carcass belonged to Number 100, a harbor seal pup whose health had rapidly deteriorated, for reasons unknown, and recently had to be euthanized. I'd never met the pup, which I was thankful for, as that made it far easier to emotionally detach from the task at hand.

Katie instructed me to suit up with a face shield and latex gloves. As she inspected each of the seal's organs, she dictated notes for me to record on a series of forms. I then photographed each organ and labeled and packaged the tissue samples. "It looks angry," she dictated as she examined the stomach, "but I don't want you to write angry in the notes." I hovered my pen above the clipboard as she considered her phrasing. "You can write, 'subjectively appears irritated.'"

By her own estimation, Katie had conducted over one hundred necropsies. She'd worked closely with Kip Temm, the consulting veterinarian for Marine Mammals of Maine, to understand what clues to look out for that might help determine a cause of death, and how specific words could help a pathologist who would later analyze the tissue samples. She'd also learned from Lynda, who had conducted countless necropsies over the past twenty years, not just on seals, but whales, dolphins, porpoises, and other marine wildlife.

"We have just one organ left," said Katie. She eyed me with apprehension. "It's the brain. This part isn't very pretty, so I want to give you fair warning if you don't want to watch." I assured her I was fine to continue, although it did make me wonder how many "fainters" they'd had to deal with over the years. As Katie prepared her tools, she lifted the seal's head and turned it toward me on the table. I took a quick breath. It was the first time I'd seen Number 100's face. The seal's large eyes, glassy and half-closed, were a startling reminder of the life that once existed within this collection of meticulously labeled organs and tissues spread across the table.

Katie picked up a metal wedge pick and a small mallet, which she used to lightly crack the seal's skull. She worked slowly, careful not to damage the fragile brain tissue. Once she'd made a sizable crack, she used her thumbs to gently pry open the skull. She then scooped out the brain in a single, fluid motion, placing the squishy blob on the table. "I'm going to take a small piece of brain for my frozen samples," she said, after inspecting the organ. "But the rest can go in there." She pointed to a large plastic container. "They like as much brain as possible." I briefly considered making a zombie joke as I carefully lowered Number 100's brain into the container, but decided against it.

I asked Katie if necropsies were more challenging the more time she'd spent with the seal. For the most part, she said, she was able to stay focused on the scientific aspects of the work, using the opportunity to learn from the animal's death. Necropsies offer insight on diseases circulating within seal populations, such as the outbreak of avian influenza in seals the prior year, or whether there are issues with water quality that impact a seal's survival. Still, sometimes, after she's done, she feels the weight of it. Some of the hardest cases have been the underdogs—the seals the staff wholeheartedly root for, against all survival odds.

Her words resonated. It wasn't long ago that a harbor seal pup had arrived at the center with the most brutal collection of slash wounds I'd ever seen. She had been attacked by a shark, but miraculously, despite the severity of her wounds, she was still alive.

Lynda had placed the seal in a quiet room at the center to recover from the trauma. The pup was alert but still when she brought me in to see her. Her head was raised slightly, turned away from us. We were both silent as I observed the seal's injuries. I'd been interested in how the size and shape of seal wounds could help scientists identify a specific shark species. The depth and jagged angles of the slashes across her back were a useful example.

But my emotional detachment from the seal evaporated as soon as she turned her head to look at me. In contrast to her shredded back, the pup's face was flawless—her large, expressive eyes accentuated by her silver metallic fur. As I locked eyes with this brutally wounded creature, something lurched deep within my gut and I felt, for a fleeting moment, the weight of her pain, fear, desperation, her tenuous existence.

After we'd returned to the office, I asked Lynda if she thought the pup would survive. "It's hard to say at this point," she said, "but it's a long shot. She's still in shock, but we should know more tomorrow."

The following morning, as I waited in line for a coffee, I sent Lynda a message to ask how the seal was faring. It was just after seven thirty, but I knew she'd already been up for hours. She wrote back immediately: *She didn't make it. I'm sorry.*

Of course, on a rational level, this wasn't a surprise. What creature could possibly survive those injuries? And yet, in the middle of the café, my eyes filled with tears. I grabbed a napkin, wholly unprepared for such an emotional response to a creature I'd spent no more than three minutes with. Why that seal, why that moment? Maybe I was still trying to make sense of the chaos, even after nearly a year of seeing that nature wasn't interested in what humans think is right or wrong. Maybe I just wanted to root for an underdog.

The team wouldn't receive the results from the necropsy for months. Number 100's case was catalogued with many others, including Number 39, the pup that had recently been released, and Number 80, the Bailey Island pup that didn't survive, as cases of "human interaction."

According to Kristina Cammen, associate professor of marine mammal science at the University of Maine, when it comes to human interactions with wildlife, there tends to be a consistent storyline. First, we exploit the species, depleting its population. Then we try to save it, attempting various conservation efforts to bring it back. Then, in cases where a species successfully recovers, we face two options. The first is to leave the conservation measures in place and allow the species to hit its ecological carrying capacity—where natural (meaning nonhuman) controls limit the continued growth of a population. For seals, this could take the form of predators like sharks or killer whales, disease outbreaks, or limited food resources. The second option is to "manage" the species, either by eliminating conservation measures or potentially even reversing them, such as through hunting. The challenge with this second method, as Kristina and her colleagues explain in a 2019 article published in *Ecosphere*, is the risk of entering into a continuous loop of depletion, conservation, and recovery.

I wondered what would happen if policymakers and wildlife officials determined that seal populations did need to be actively managed—if Congress voted to lift the provisions of the Marine Mammal Protection Act as it exists today to enable some sort of seal population control, kicking off this continuous loop. Putting a pin in the huge ethical, logistical, and ecological questions for a moment, in this scenario, what would be the impacts on the overall health of the seal population?

I'd asked Kristina this question on a video call. As a marine scientist who was trained as a molecular ecologist, she had a unique understanding of the issue. "Presumably that means you're going to have even less genetic diversity that comes through on the other side," she said. "What we generally think about is that lower diversity makes animals less fit and more susceptible to disease." The population would also have less capacity for future adaptations, which is concerning in the midst of the significant environmental changes spurred by the climate crisis. "With an unknown future, ideally, you hedge your bets," said Kristina. "You don't know if genetic variant A or B is going to be best for a future scenario that we can or cannot predict. So as long as we have both A and B in our population, that makes us better off."

But there is still a lot of uncertainty about the future population trajectories of seals and other recovering species, she added. She used northern elephant seals as an example. After extensive hunting in the nineteenth century nearly wiped out the species, the population of northern elephant seals has grown from just *twenty* individuals to more than 200,000 today. Kristina explained that it's possible the population is exceedingly vulnerable to some new pathogen it hasn't yet been exposed to, but as of now, that vulnerability hasn't been observed. "Seals in particular seem to have this really amazing ability to recover," said Kristina. "They're resilient."

ONE OF THE MOST poignant examples of conflicts surrounding wildlife management is the resurgence of gray wolves in the West. Wolves were once hunted to near-extinction in the U.S. But in 1995, the U.S. Fish and Wildlife Service reintroduced gray wolves into Yellowstone National Park and remote areas in central Idaho. As their population expanded, they began to alter the ecosystem in remarkable ways.

In the absence of wolves, the population of elk, their preferred prey, skyrocketed. The elk began feeding year-round on aspen, willow, and cottonwood, killing off and suppressing the vegetation within the ecosystem. But when the wolves returned, the elk changed their foraging behavior. They became much more vigilant, forced to keep moving during the winter to avoid wolf predation. As a result, the vegetation began to recover. As willow stands returned, so did the beavers that relied on this food resource. The beavers spread, building dams and ponds that provided habitats for fish species. The recovering trees created habitats for songbirds. Meanwhile, scavengers like eagles, ravens, and coyotes, along with black and grizzly bears, benefited from increased wolf predation, as they fed on the remains of elk the wolves left behind.

But over time, wolves expanded beyond the protected areas in which they were released. Some began preying on livestock outside the boundaries of the park, impacting the livelihoods of neighboring ranchers.

As wolf populations recovered, they eventually reached what scientists determined to be a sustainable population threshold. Since they weren't considered to be endangered, they were "de-listed," which meant they no longer qualified for federal wildlife protections. In response, Montana, Idaho, and Wyoming opened wolf hunting seasons as a means of controlling the population and limiting predation on livestock.

In 2008, it was discovered that wolves were breeding in the northeastern part of Washington. The news was celebrated by conservationists, but infuriated local ranchers. State officials worked to develop a Wolf Management Plan, and weighed whether or not to create a wolf hunting season for the state. Tensions escalated. Washington's Department of Fish and Wildlife convened an advisory group, which was made up of livestock owners, conservationists, residents, local officials, and other stakeholders with competing interests, but they struggled to reach any kind of consensus. In the end, the state hired an independent organization, the Center for Conservation Peacebuilding, to help resolve the conflicts.

Francine Madden, the organization's executive director, spent three years working with stakeholders in Washington State to understand the depth of the issues, and to build consensus around possible solutions. By the end of her contract, conservationists and wolf advocates had agreed on the need for a lethal removal option in cases where wolves were repeatedly targeting livestock. Ranchers and other proponents of lethal removal had agreed on the need for nonlethal methods to deter wolves from preying on livestock. A number of participants were interviewed by various media outlets after the contract ended, but most seemed to have a hard time describing exactly what Madden *did* to help opposing sides reach a consensus. Some even used terms like "magic."

I'd first learned about Madden's work from Cynthia Wigren during my visit to the Atlantic White Shark Conservancy in Cape Cod. Cynthia and Megan had participated in a training offered by the Center for Conservation Peacebuilding to better understand how to address conflicts around white sharks, seals, and humans. As Cynthia spoke about the frustrations she'd heard from different communities in Cape Cod over rising seal populations, she referenced what she'd learned in the training. Of particular importance was taking the time to identify the underlying issues that might be driving the wildlife conflict.

In a 2014 perspective article Madden published in the journal *Biological Conservation*, she and her coauthor argue that in some of the more complex scenarios, wildlife conflicts can represent a manifestation of deep-rooted social conflicts. In the case of wolves in Washington, while the issue at hand was about wolf management, the underlying conflicts between stakeholders ran far deeper. They represented rural and urban divides, political divides between Seattle progressives and conservative residents of eastern Washington, divisions between government officials and citizens, hunters and conservationists. Finding a solution to address the wolf situation required understanding and addressing the underlying needs and cultural divisions between these groups, while building trust and empathy for opposing viewpoints.

"You have to recognize people are going to have different opinions," said Cynthia. "And you need to be empathetic, to listen and try to understand rather than shutting people out who don't agree with you."

Cynthia recently joined the first-ever international conference focused on human-wildlife conflict and coexistence, organized by the International Union for Conservation of Nature. After listening to examples of wildlife conflicts from all over the world, she said that one of her biggest takeaways was that conflicts don't just one day stop. "It's about how can we best manage it and move through it," she said. "And that does come down to a lot of communication."

RESPONDING TO WILDLIFE CONFLICTS has tended to be a reactive process. But many are calling for a more proactive approach—to identify conflicts earlier, before they deepen to the point of undermining conservation goals.

In some cases, scientists are even finding ways to predict future conflicts before they happen. One example of this is occurring in Montana's Glacier National Park, where a rebounding population of grizzly bears has begun to move beyond the boundaries of its recovery zone. The elusive nature of grizzlies makes it difficult for wildlife managers to track them. But scientists have found a way to predict their movements by mapping one of their most important food resources—huckleberries.

A team of scientists led by Tabitha Graves, a U.S. Geological Survey research ecologist, has trained a computer to map Glacier's huckleberry plants with remarkable accuracy by combining field data with satellite imagery and machine-learning. Graves and her colleagues are also studying shifts in the distribution of these plants in response to factors like wildfires and climate change. Warming temperatures, declines in wild pollinators, and varying levels of rainfall, for example, can alter the size and availability of huckleberries, as well as the timing of their fruiting. By understanding where huckleberries are now, and where and when they may be fruiting in the future, the data can be used to predict bear movements, and to map areas where these movements may overlap with human populations.

AS CYNTHIA SAID, "SCIENTISTS often get into this work because they want to study a certain animal they like, but it ends up being so much more about people."

Writers, too, I thought. I'd spent a year trying to understand our complex relationship with seals by exploring the lives of these charismatic marine mammals and the humans who interact with them, only to come to the realization that on some level, the conflicts aren't about seals at all.

Humans have an evolutionary connection to the ocean. As Rachel Carson wrote in *The Sea Around Us*, "When they went ashore, the animals that took up a land life carried with them a part of the sea in their bodies, a heritage which they passed on to their children and which even today links each animal with its origin in the ancient sea."

Perhaps some of the anger toward seals has more to do with the perception that the lives of these animals are being prioritized over the lives of humans. Fishing has an important place in our heritage, and at a global scale, we rely on fisheries to help feed the planet. In my view, seals serve as a particularly convenient scapegoat. If seals were the only problem, then the solution, in theory, would be quite simple. Kill the seals, bring back the fish. But our fisheries are threatened by issues that are far more complex than hungry, hungry seals.

During the months I'd spent visiting the charismatic seal pups at Marine Mammals of Maine, I'd found it quite easy to adopt a total protectionist stance. Yet I also recognized that I'd never been personally inconvenienced by a seal. I'd never lost wages due to seals, nor felt a sense of competition with seals over a shared resource. And while I'm much more cautious about when and where I take a dip in the ocean these days, I'm not a long-distance swimmer, and I've never (successfully) surfed. I'd never felt that the lives of seals were being prioritized over my own, nor that my identity or livelihood was being impacted.

Plus, in what had felt progressively more hypocritical as my exploration of the world of seals continued, I loved seafood. And I'd begun to better understand how our insatiable demand for certain types of fish—cod, haddock, salmon, and steelhead, to name a few—in all months of the year and for the cheapest prices possible, has fueled a deeply flawed commercial fishing industry. And, as I'd come to appreciate, this industry is inextricably linked to the fate of seals.

16

The Shape-Shifter

SUMMER PASSED WITH THE speed of a bullet train. I waited for clarity. It never came. By August, the ground beneath me felt wobbly, as though I was adjusting to dry land after months at sea. Of course it had been unrealistic, even arrogant, to assume I'd reach some profound conclusion about seals, about our relationship with these charismatic marine mammals, about our ability to address wildlife conflicts that date back to antiquity. And yet, I certainly didn't expect that after completing over a year of research, I'd have so much *less* clarity than when I started.

The well-intentioned questions from family, friends, even new acquaintances, didn't help. "Why are there suddenly so many seals?" they'd ask. "Where did they come from? What do they eat?" Then, eyes widening, "And what about the sharks? Is it safe to swim in the ocean?"

I didn't feel equipped to respond. "It's complicated," I'd reply with a polite smile. Almost immediately, their eyes would glaze over and the conversation would shift. *No one* wanted to hear about complicated.

Another challenge was that I'd begun to see the seal controversy as a microcosm not just of human-wildlife conflict, but of human-human conflict. The world felt volatile, fractured, more divided and disconnected than I'd ever known it to be. On some level, maybe I thought if I could make sense of our relationship with seals, make sense of the ways in which humans were divided in their perspectives on seals, then I could make sense of all the rest (*gestures to grand horror of modern life*). In the process of learning more about these creatures, and getting to know the humans who most closely interact with them, I'd hoped to discover some sort of meaning, some new way to make sense of my own place in this world. Through diligent, focused research, I'd simply compile the data, shake it up, and pour out a neat glass of clarity, bursting with botanicals.

Back in the spring, during my visit with members of the Puyallup Tribe, Ramona Bennett asked me what should have been a simple question: What was my story about? I shared with her a rambling version of "it's complicated"—the disappointing truth. Ramona looked at me thoughtfully, then nodded. "The more you know, the more you know you don't know," she said. I looked up at her, surprised. "Yes. That's it *exactly*."

Yet I still felt as though I'd somehow failed in my mission. We're trained from an early age to view the world in the binary—right and wrong, black and white. Embracing uncertainty, that messy middle ground, was exhausting. I was exhausted. But on some level, maybe it wasn't such a bad thing.

In her book, *Uncertain,* author Maggie Jackson argues that uncertainty is not a flaw to be avoided, but rather a skill set to cultivate—something that can help us to navigate complex crises. "Ambivalence inspires a more subtle understanding of the problem and actions better calibrated to the situation," she wrote. In a polarized society, uncertainty can help to bridge divides. To be less certain, less knowledgeable, means that you are open to new ways of thinking, seeing, and feeling. You can innovate. You can adapt. I was trying to accept this, even as my mind swirled with all the competing facts, beliefs, and priorities I'd collected over the course of the year.

AT THE END OF August, I joined three friends for a long weekend in Brooksville, Maine—a seaside town on the coast of Penobscot Bay. The weekend was straight-up magic, filled with morning hikes, afternoon swims, feasts of fresh lobster, scallops, and corn doused in melted butter. In the evenings, we sat by the fire, warmed by embers and wine, while our dogs slept soundly beside us. Bats flitted through the darkness, chasing after their insect suppers, as we polished off blueberry crumble and key lime pie.

It wasn't lost on us how lucky we were to be there—to unplug from the chaos of the world and immerse ourselves in nature. We walked amongst the rockweed beds as ten-foot tides receded dramatically from the shore, revealing the hidden worlds of the semi-aquatic life beneath. Horseshoe crabs burrowed into mudflats, terns dive bombed into the harbor, great blue herons passed overhead, their long wings balancing in the wind. Wildflowers grew in thick patches along the dirt road, brushstrokes of cheerful yellows and purples framing the landscape.

One afternoon, we decided to check out a nearby rock where seals were known to gather. As we walked from the house to the harbor, I was reminded of the children's book *One Morning in Maine* by Robert McCloskey (of *Blueberries for Sal* fame), which happened to be set in Brooksville. In the book, young Sal has a loose tooth. After breakfast, she walks to the harbor to help her father dig for clams. Along the way, she shows off her wiggly tooth to an osprey and a loon, but neither pays much attention. Sal ". . . was just about to go her own way when a seal poked his head up out of the water. 'I have a loose tooth!' Sal said to the seal, and the seal, being just as curious as most seals, swam nearer to have a good look."

We've long been drawn to seals' curiosity. Perhaps it's because it makes us feel interesting, but I believe it's deeper than that. We identify with it. Humans are hardwired to be curious, to search for meaning, to try to make sense of the world around us. As ecologist James Estes wrote, "We strive to learn because that is what humans have always done. Learning and understanding are a central essence of humanity."

My friends and I shoved off from shore in our four kayaks. Not far from the dock was a small island with tall white pines rising from the edge of the shore. From the highest branches, two bald eagles peered down at us as we paddled by. But when we reached the other side of the island, instead of a rock teeming with seals, we saw only the flat surface of the water. We'd misjudged the tides, and the rock was fully submerged. It would be hours before the water receded enough for it to resurface, for the seals to return.

Disappointed, we floated silently, four red and yellow kayaks drifting along their own paths, bobbing in the surf. I wondered where the seals went when the tide came in. I imagined they used the time to swim offshore to hunt for fish, but perhaps they had a back-up resting spot not far from here. I supposed I'd never know.

We believe we understand seals because we share, at times, this small sliver of coastline where our physical worlds collide. For brief moments, we observe them, as they observe us. We laugh at their clumsy, awkward movements, their banana-shaped poses as they dry their heads and hind flippers in the salty air. We have a small window into their land lives, but their underwater lives remain a virtual mystery. Beneath the surface, seals transform from giant russet potatoes sunbathing on a rock into sleek torpedoes—their flippers propelling them through the water at tremendous speeds, their movements graceful, agile. With the help of satellite tracking and other research technologies, we've gathered some useful data on seals' underwater movements and behaviors. But the truth is, we've barely scratched the surface.

Even Lynda, with her decades of seal research experience and animal interactions, has acknowledged how little we know about the lives of seals. "There's so much that goes on under the surface of the ocean that we can never see," she told me. "There's so much more to be discovered and explored. We have only a snapshot of their lives."

I don't know how long the four of us humans floated there in the water, each of us lost in our own quiet reflections. Eventually, a loud splash startled us back to the moment. As we looked around, searching for the source, a dark, shiny head popped up between our kayaks, then dove back under, smacking the water with its hind flippers. Soon, another head emerged, and then another. I wondered if they'd been swimming beneath us this whole time. Perhaps they were drawn to the surface simply to take a breath. Or perhaps, like us, they were curious.

Soon after, more than half a dozen harbor seals were bobbing like soda bottles a short distance from our boats. They spun slowly in the water as they looked at each of us in turn, their heart-shaped nostrils flaring.

The anthropologist Loren Eiseley once wrote, "One does not meet oneself until one catches the reflection from an eye other than human." What we see when we look at a seal reveals far more about us than it does about the liquid-eyed creature staring back. It's impossible to disentangle our perceptions of seals from our own lives and experiences, beliefs and biases. We'll never know what these creatures think and feel, the depth of their wild existence.

The more time I'd spent watching and thinking about seals, the more I'd considered my own role in nature, my tolerance for inconvenience, the power I hold to control, protect, destroy, respect other species. The more I studied seals, the more I learned about what it means to be human.

We each shape the stories of seals in the way that suits us, but there's inherent fragility in that. Our shifting perceptions of these creatures—as competitors, symbols, guides, beloved companions—impact their very survival.

ON A LONELY ROCK on the edge of a distant island in the North Atlantic, there is a statue of a woman. She is made of steel and bronze and holds in her hand a sealskin. It gathers in folds at her feet and drapes across one of her legs, as though she was frozen midway through stepping out of it.

The statue, whose name is Kópakonan, or Seal Woman, was raised in 2014 in the northern part of the Faroe Islands, sculpted by a Faroese artist. Months later, a winter storm ravaged the coast, and a forty-foot wave swept over the statue, but she was unharmed. After all, she's part seal.

Kópakonan is a selkie—a mythological being capable of shifting between seal and human forms. Rooted in Nordic and Celtic oral history and culture, selkie stories each have slight variations. But in the Faroe Islands, Kópakonan is the selkie tale best known.

As the legend goes, every year, selkies gather in a cave on the Faroese island of Kalsoy. They emerge from the sea, shed their skins, and live as humans for a single night. On hearing the story of the selkies, a farmer from the nearby village of Mikladalur decides to see it for himself.

As the sun sets, the farmer hides behind a rock near the cave. Soon after, a group of seals appear in the water, their heads rising above the surf as they make their way to shore. One by one, the selkies remove their skins, revealing their human forms. They dance and sing, beginning their evening of revelry. The farmer, still hidden behind a rock, is entranced by one of the selkies who has morphed into a beautiful young woman. Overcome with desire, he sneaks down to the beach to steal her sealskin.

At dawn, the selkies gather again on the beach to retrieve their skins and return to the sea. Only the beautiful selkie woman, Kópakonan, remains. It is then that the farmer reveals himself, holding her sealskin in his arms. She has no choice but to return with him to Mikladalur. The farmer locks her sealskin in a chest, careful to keep the key always in his belt.

Eventually, they marry and have several children. While Kópakonan does her best to adapt to life as a human, she is often seen staring longingly at the sea. One day, the farmer, while out fishing in his boat, realizes that he's forgotten the key to the chest. He rows to shore as fast as he can, fearing the worst. When he reaches the house, he finds his children sitting silently at the kitchen table. Their mother has returned to the sea.

Time passes until one day, the men from Mikladalur plan a seal hunt at the cave near the village. The night before the hunt, Kópakonan appears in the farmer's dream. She begs him to spare her selkie husband and two sons. But on the day of the hunt, the farmer, devastated by the loss of his wife, and fueled by jealousy for her selkie husband, kills all three.

That evening, the villagers hold a feast to celebrate the successful hunt. As the meal is being prepared, the door bursts open, revealing Kópakonan. When she sees her mate and two pups, their heads and flippers arranged on platters, she screams and curses the villagers. She foretells of men from Mikladalur who will lose their lives at sea. As many will die by drowning or falling from the highest cliffs, she says, as can link arms and circle the whole island of Kalsoy.

HUMANS HOLD IMMENSE POWER to shape and control the lives of other creatures, but selkie mythology demonstrates the futility of those efforts. If there is even the tiniest window of opportunity to do so, nature will return to its ways. But that doesn't absolve us of responsibility.

In the Faroe Islands, where the statue of Kópakonan stands, a small breeding colony of gray seals has managed to survive despite over one thousand years of intensive seal hunting. To escape their most powerful predators, at some point seals abandoned their coastal haul-out sites and started taking to ocean caves to breed, pup, and molt. Some of these caves can only be accessed through underwater entrances, and are therefore impossible for humans to reach. The animals adapted to survive.

We made every effort to eliminate seals from American waters. The fact that we came so close to succeeding makes it all the more remarkable that these animals have managed to recover, just fifty years later. They adapted to our presence. Now, it remains to be seen whether we can adapt to theirs.

Perhaps the question shouldn't be whether or how to control nature, but how to control ourselves. We're better positioned than any other creature on earth to adapt our ways to coexist with the wild world. But I'd come to realize that to achieve lasting conservation goals, we'd need to confront the deeper conflicts that exist among us.

THE MARINE MAMMAL PROTECTION Act was the first legislation in the U.S. whose primary objective was to support the health of ecosystems, as opposed to the health of individual species. From the outset, it recognized the interconnectedness of marine life.

"It's about our relationships," said Dwayne Tomah, the Passamaquoddy Tribe's language keeper, during my visit to Sipayik. I had asked him about the tribe's relationship with seals, but Dwayne had just looked at me, seemingly confused. "The relationship we have with the water, the relationship we have with the fire, the relationship we have with the animals, the relationship we have with human beings. It's all connected."

Looking back, I better understand what he was telling me. I had been so fixated on these individual species, when seals were just one piece of a far larger, multidimensional puzzle, just as humans were another piece. The only way to make sense of these issues was to look at the health of the whole, not the health of the individual. Scientists and government officials often talk about "managing" resources, whether that's seals, fish, trees, rivers, or any other natural systems. But language is everything. For the Passamaquoddy, as has been the case for so many Indigenous people, the point isn't to manage a resource, but to maintain a relationship. The phrasing allows for usage—hunting, fishing, foraging—but that usage must be sustainable, and goes hand in hand with preservation and reciprocity.

As with so many environmental and political questions, we're desperate for wildlife conservation to be simple, straightforward. But those of us who claim to be nature purists, advocating for total protections for wildlife, are often those with the least amount to lose. Over the course of my year of reporting, I'd come to better understand the impacts of seals on coastal communities, both real and perceived, as well as the depth of the divisions that exist between the humans who interact with them most frequently. Honestly, it seems pretty hokey (or perhaps all too obvious) to say that the secret to lasting conservation policy comes down to empathy, to finding middle ground, to strengthening human relationships. But I'd seen firsthand the value of the partnerships that have developed, between fishermen and conservationists, scientists and surfers, Native and non-Native resource maintainers.

Even Lynda, Maine's "seal saver," relies on fishermen to support her work. Perhaps the loudest anti-seal voices have been those of fishermen, but fishermen are also the ones calling the Marine

Mammals of Maine hotline to report stranded seals in every season. And when Lynda calls in a favor, they show up for her. She's bridged the divides with fishermen in part because she is one. She respects their values, their livelihoods, their beliefs, even if they don't always agree on everything.

Months earlier, she'd told me about a friend of hers, Merle Gilliam, who died in 2022 at the age of ninety-four. Gilliam often teased Lynda about how he used to shoot seals when he was younger. He'd fished for most of his life, since well before the Marine Mammal Protection Act. He grew up at a time when shooting seals wasn't just common practice, it was encouraged by the state. For her part, Lynda took the teasing in stride. She'd roll her eyes and threaten to name a rescued seal after him, or to hang a large plaque bearing his name above the rehab center. And she honestly enjoyed hearing about his experiences on the water and the waves of ecological changes he'd witnessed.

"It's not about convincing people that they're right or wrong if it's what they believe," Lynda told me. "A hundred years from now, it might go back to what it used to be like for seals, I don't know. But we have to understand the history. We have to be open to those conversations."

Embracing nuance isn't always easy. But throughout my reporting, I bore witness to many of the systemic challenges within the industry at the heart of human-seal conflicts: fisheries. I've come to appreciate my own role as a seafood consumer in driving demand for overfished species. As one Cape Cod fisherman phrased it—it's time we stop telling the ocean what we want to eat, and instead eat what the ocean provides. The sea is a shared resource. We all own it. Not just us humans, but seals, whales, and the fish at the epicenter of these battles. I'm learning, slowly, how to make changes—to ask better questions, to understand where my fish is coming from—not just the region, but the boat, the gear being used—to start to understand what it might mean to eat what the ocean provides.

It's unlikely that human conflicts with seals and other wild creatures will ever go away. If anything, they'll likely increase over time as our population expands and we further degrade and fragment natural habitats. The question will be how we manage those conflicts, and whether we'll take the time to identify and address the deeper issues at the heart of them.

ON OUR FINAL EVENING in Brooksville, my friends and I returned to the dock as the sun was setting. The tide was about halfway out, and we could hear the seals in the distance, splashing and croaking like bullfrogs. I borrowed a friend's paddleboard and binoculars, and slowly, quietly, paddled toward them. About two hundred feet from the rock, I stopped and dropped to my knees on the board. I kept still, at a distance, to make sure I didn't interfere with their rest, or worse, startle them into flushing off the rock. Through my binoculars, I counted sixteen seals hauled out. All sixteen were staring, curiously, in my direction.

How similar, the ways in which we identify with seals. And yet there is something so exotic, so unknowable about them. They're as wild as animals in a human-dominated world can be. It's hard to accept there are some things we can't ever know. But there's also hope and purpose in the not knowing. Mystery and curiosity power our pursuit of knowledge. In the meantime, we imagine, we wonder, and marvel.

The sun was halfway beneath the horizon by the time I turned and paddled back to the dock. The cacophony of burps and croaks grew fainter as I neared the shore. I smiled, heartened by the resilience of these wild creatures. Seals will endure, if we let them.

ACKNOWLEDGMENTS

I'M GRATEFUL TO THE many generous individuals mentioned in this book who gave of their time, insights, and expertise. I'm especially indebted to those who welcomed me into their worlds and helped me to understand the depth and nuance of this story.

An ocean of thanks to the Algonquin team. To my brilliant editor, Maddie Jones—your belief in this story, warm encouragement, and ability to nudge me out of the seaweed to gaze at the kelp forest now and again helped to make this book what it is. Thank you to Catherine Schott and Steve Godwin for the stunning cover art and interior design; Martha Cipolla for your invaluable copyedits; Mira Park for your eagle-eyed review, Brenna Franzitta and Brunson Hoole for managing the millions of pieces required to bring this book to life; and Marisol Salaman and Kara Brammer for their support in getting it into the hands of readers. And thank you, Margot Hart, for the magnificent maps.

To my agent, Michelle Tessler—thank you for seeing potential in this book and finding it the greatest home at Algonquin.

I'm grateful to the Alfred P. Sloan Foundation and its program to support the public understanding of science and technology, which gave me the confidence to advocate for this story along with the resources to invest in its research. I'm also grateful to *Down*

East Magazine for publishing an earlier version of one of the stories in this book.

A sea of gratitude to my early readers—my mom, Tina Morris, and my dear pal, Julia Coit—your enthusiasm, curiosity, and insightful edits were a lifeline, particularly in moments of doubt.

This book would not exist if it wasn't for Sy Montgomery, who not only convinced me to write it but supported me each step of the way. I couldn't ask for a more passionate and inspiring mentor and friend, nor a better roommate while tracking wild dogs in the Khao Yai jungle.

There are a number of people who shared their knowledge and offered valuable research support who weren't mentioned in the book, including: John Calambokidis, Demian Chapman, Candace Cochrane, Rob DiGiovanni, Niaz Dorry, Brenna Frasier, Charlie Innis, Emily Jones, Katie McConnell, John Mandelman, Nancy Milburn, Dennis Minty, Nicole Morrill, Larry Morris, Jennifer Nicholson, Wendy Puryear, Sue Ramin, Stan Rullman, Pam Snyder, Anne-Seymour St. John, Susan St. John, Phil Thorson, Dave Testaverde, and Feini Yin. I appreciate you all.

To Lynda Doughty, Dominique Walk, Katy Green, Katie (Gilbert) Jenner, Lexi Wright, and the rest of the team at Marine Mammals of Maine—thank you for allowing me to inundate you with four seasons' worth of seal questions, for your trust, for the laughter and the tears, and for your passion, heart, and grit. I'm inspired by you all.

To my wonderful family and friends who provided encouragement and inspiration along the way—I couldn't have done this without you. A special thanks to Hilary Langer and Kathie Duncan—your cheerleading, check-ins, and surprise food deliveries meant more to me during the foggiest days of book writing than you'll ever know.

Marlow Kendall—your quiet patience and humor as I waxed

eloquent about pinnipeds and fisheries science these past two years was a gift. Thank you for your unwavering confidence in me, for challenging me to think critically, and for showing me new ways to see and understand nature.

To Kate Hedlund-Groden, Lindsey Coit, and Julia Coit—my "seal squad." You lifted me up in the darkest hours, and I couldn't have done this without you. Thank you for your patience as I turned multiple years of wilderness adventures into seal scouting missions, for inspiring me with your own courageous journeys, and for the bottomless blubber belly laughs that kept me whole and breathing.

Lastly, thanks to my home office mate, Quinn the Cat, for her warm companionship, and to the two worst behaved and *best ever* land pups, Lucy and Lou, whose big, liquid eyes and desire to push every possible boundary in pursuit of treats proved to be of unique relevance to this story.

BIBLIOGRAPHY

INTRODUCTION

Flaherty, Nora. "Scientist: Increase in Seal Population Likely Attracting More Sharks to Maine Waters." *Maine Public* (July 28, 2020). https://www.mainepublic.org/environment-and-outdoors/2020-07-28/scientist-increase-in-seal-population-likely-attracting-more-sharks-to-maine-waters.

Graham, Gillian, Kelley Bouchard, and Reuban Schafir. "Three Reported Shark Sightings Put Maine Marine Patrol on High Alert Following Fatal Attack." *Portland Press Herald* (July 29, 2020). https://www.pressherald.com/2020/07/29/coastal-patrols-continue-after-maines-first-fatal-shark-attack.

Lelli, Barbara, David E. Harris, and Abouel-Makarim Aboueissa. "Seal Bounties in Maine and Massachusetts, 1888 to 1962." *Northeastern Naturalist* 16, no. 2 (2009): 239–54. https://doi.org/10.1656/045.016.0206.

Maine Department of Marine Resources. "Update on Deadly Shark Attack." Video, WGME CBS 13 News, Portland, ME (July 2020). https://www.facebook.com/watch/live/?ref=watch_permalink&v=215922843052070.

Miles, Kathryn. "Shark Attacks in Maine Were Unthinkable—until Bailey Island." *Down East Magazine* (June 2021). https://downeast.com/land-wildlife/shark-attacks-in-maine-were-unthinkable-until-last-summer.

Smalley, Suzanne. "After Fatal Great White Shark Attack in Maine, Debate Intensifies over Culling Seals." *Yahoo! News* (August 8, 2020).

https://www.yahoo.com/news/summer-of-the-shark-after-maine-attack-shark-experts-say-the-beasts-are-here-to-stay-020602175.html.

Trotta, Daniel. "Maine Killing Shows Great White Sharks Are Back, Attracted by Seals." *Reuters* (July 29, 2020). https://www.reuters.com/article/world/maine-killing-shows-great-white-sharks-are-back-attracted-by-seals-idUSKCN24U2M6.

U.S. Congress. "Marine Mammal Protection Act of 1972," amended 2019. NOAA Fisheries. https://www.fisheries.noaa.gov/national/marine-mammalprotection/marine-mammal-protection-act.

Yurk, Valerie. "Rare Shark Attack in Maine May Be Linked to Marine Protection Efforts." *The Guardian* (July 30, 2020). https://www.theguardian.com/environment/2020/jul/30/maine-shark-attack-marine-protection-efforts.

CHAPTER 1: A SEAL WITH SOMETHING TO SAY

Bennett, Ruth. "Mom, Is That You? Seals Show Family Recall." *Science News* (March 24, 2003). https://www.sciencenews.org/article/mom-you-seals-show-family-recall.

Boston University. "Hoover the Talking Seal." Audio and Video, Guenther Speech Neuroscience Lab. https://sites.bu.edu/guentherlab/miscellaneous-videos-and-oddities/hoover-the-talking-seal. Accessed September 19, 2023.

Cammen, Kristina M., Douglas B. Rasher, and Robert S. Steneck. "Predator Recovery, Shifting Baselines, and the Adaptive Management Challenges They Create." *Ecosphere* 10, no. 2 (2019): e02579. https://doi.org/10.1002/ecs2.2579.

Currier, Patricia A. "Hoover Will Talk No More; a Delight to Thousands, Aquarium Seal Dies at Age 14." *The Boston Globe* (July 26, 1985).

Gould, Stephen. "A Biological Homage to Mickey Mouse." *Ecotone* 4 no. 1 (2008): 333–340. https://doi.org/10.1353/ect.2008.0045.

Hall, Danielle. "Seals, Sea Lions, and Walruses." Smithsonian Ocean (last modified January 2021). https://ocean.si.edu/ocean-life/marine-mammals/seals-sea-lions-and-walruses.

Haverkamp, Holland, Hsiao-Yun Chang, Emma Newcomb, et al. "A Retrospective Socio-Ecological Analysis of Seal Strandings in the Gulf of Maine." *Marine Mammal Science* 39, no. 1 (2022): 232–50. https://doi.org/10.1111/mms.12975.

Hiss, Anthony. "Hoover." *The New Yorker* (December 27, 1982). https://www.newyorker.com/magazine/1983/01/03/hoover.

Lambert, Robert A. "Grey Seals: To Cull or Not to Cull?" *History Today* 51, no. 6 (June 2001): 30–32.

Lambert, Robert A. "The Grey Seal in Britain: A Twentieth Century History of a Nature Conservation Success." *Environment and History* 8, no. 4 (November 2002): 449–74. https://doi.org/10.3197/0967340021 29342738.

Lambert, Robert A. "Environmental History and Conservation Conflicts." In *Conflicts in Conservation: Navigating towards Solutions*, edited by Stephen M. Redpath, R.J. Gutierrez, Kevin A. Wood, and Juliette C. Young. Cambridge: Cambridge University Press, 2015.

Liu, Xiaodong, Suzanne Rønhøj Schjøtt, Sandra M. Granquist, et al. "Origin and Exansion of the World's Most Widespread Pinniped: Range-Wide Population Genomics of the Harbour Seal (*Phoca vitulina*)." *Molecular Ecology* 31, no. 6 (2022): 1,682–99. https://doi.org/10.1111/mec.16365.

Magera, Anna M., Joanna E. Mills Flemming, Kristin Kaschner, Line B. Christensen, and Heike K. Lotze. "Recovery Trends in Marine Mammal Populations." *PLoS ONE* 8, no. 10 (2013): e77908. https://doi.org/10.1371/journal.pone.0077908.

"Marine Mammals of Maine 2022 Annual Report." Marine Mammals of Maine. https://www.mmome.org/wp-content/uploads/2023/05/2022-MMOME-Annual-Report.pdf-4.pdf. Accessed October 3, 2023.

Mooallem, Jon. *Wild Ones*. Penguin Books, 2014.

Mowat, Farley. *Sea of Slaughter*. Atlantic Monthly Press, 1984.

"Gray Seal." NOAA Fisheries. 2022. Last modified August 6. https://www.fisheries.noaa.gov/species/gray-seal.

"Harbor Seal." NOAA Fisheries. 2022. Last modified April 19. https://www.fisheries.noaa.gov/species/harbor-seal.

Swallow, Alice Dunning. *Hoover the Seal, and George*. Freeport Village Press, 2001.

Treadwell, David. "The (Talking) Seal of Approval." *The Times Record* (January 31, 2020). https://www.pressherald.com/2020/01/31/david-treadwell-the-talking-seal-of-approval.

U.S. Congress. "Marine Mammal Protection Act."

Waring, Gordon T., James R. Gilbert, Dana Belden, Amy Van Atten, and Robert A. DiGiovanni Jr. "A Review of the Status of Harbour Seals (*Phoca vitulina*) in the Northeast United States of America." *NAMMCO Scientific Publications* 8 (September 2010): 191. https://doi.org/10.7557/3.2685.

Wood, Stephanie A., Kimberly T. Murray, Elizabeth Josephson, and James Gilbert. "Rates of Increase in Gray Seal (*Halichoerus grypus atlantica*) Pupping at Recolonized Sites in the United States, 1988–2019." *Journal of Mammalogy* 101, no. 1 (2019): 121–28. https://doi.org/10.1093/jmammal/gyz184.

CHAPTER 2: THE TRAVELING SEAL

Allen, Byron, host. "Andre the Seal | Real People." Video, NBC (January 9, 1980). https://www.youtube.com/watch?v=QnOTWo9CMmA.

Associated Press. "Harry Goodridge, 74, Seal Trainer in Maine." *The New York Times* (April 7, 1990).

Australian Antarctic Program. "Elephant Seal." Australian Government, 2018. Last modified March 20. https://www.antarctica.gov.au/about-antarctica/animals/seals/elephant-seal.

Barash, David P. "Why Did Humans Evolve to Be So Fascinated with Other Animals?" *Aeon* (May 13, 2014). https://aeon.co/essays/why-did-humans-evolve-to-be-so-fascinated-with-other-animals.

Clark, Edie. "Andre the Seal | 25 Years with Andre." *Yankee Magazine* (November 11, 1986). https://newengland.com/travel/maine/andre-the-seal.

Clayton, Lauralee. "Wisconsin Schoolchildren Start 'Fan Club' for Andre the Seal." *The Camden Herald* (1985).

"A Community Comes Together for a Gray Seal." NOAA Fisheries (September 15, 2023). https://www.fisheries.noaa.gov/feature-story/community-comes-together-gray-seal.

Demecillo, Maron. "I 'Woof' You: How Pet Pictures Influence Online Dating Selection." Thesis, MacEwan University (August 25, 2023). https://journals.macewan.ca/studentresearch/article/view/2665.

Department of Natural Resources and Environment Tasmania. "Wildlife Management | Southern Elephant Seal." Tasmanian Government. 2023.

Last modified July 25. https://nre.tas.gov.au/wildlife-management/fauna-of-tasmania/mammals/seals/southern-elephant-seal.

Dietz, Lew. "Maine's Harbor Seals." *Outdoors Maine* (1961).

Goodridge, Harry, and Lew Dietz. *A Seal Called Andre*. Praeger, 1975.

Grove, Lloyd. "The Old Seal and the Man." *Washington Post* (April 24, 1985).

Hall. "Seals, Sea Lions, and Walruses."

Harris, Kim, director. "The Seal Who Came Home." Video, PBS, August 6, 2014.

Henley, Jon. "Norway Was Right to Put Down Freya the Walrus, Prime Minister Says." *The Guardian* (August 15, 2022). https://www.theguardian.com/world/2022/aug/15/norway-right-put-down-freya-walrus-prime-minister-says-jonas-gahr-store.

Herzog, Hal. *Some We Love, Some We Hate, Some We Eat: Why It's so Hard to Think Straight about Animals*. Harper Perennial, 2011.

Horowitz, Jason. "A Famous Walrus Is Killed, and Norwegians Are Divided." *The New York Times* (August 19, 2022). https://www.nytimes.com/2022/08/19/world/europe/norway-walrus-freya-killed.html.

Lange, Ariane. "Whatever Happened to the Seal from 'Andre'?" *BuzzFeed News* (November 1, 2014). https://www.buzzfeednews.com/article/arianelange/famous-sea-lion.

Lelli. "Seal Bounties in Maine and Massachusetts, 1888 to 1962."

Louv, Richard. *Our Wild Calling*. Algonquin Books, 2019.

McConnell, B. J., C. Chambers, and M. A. Fedak. "Foraging Ecology of Southern Elephant Seals in Relation to the Bathymetry and Productivity of the Southern Ocean." *Antarctic Science* 4, no. 4 (December 1992): 393–98. https://doi.org/10.1017/s0954102092000580.

Miller, George T., director. *Andre*. Video, Paramount Pictures, 1994.

Moench, Mallory. "The Internet's Newest Sensation Is Neil the Seal from Tasmania." *Time* (December 22, 2023). https://time.com/6550164/neil-the-seal-tasmania-background-instagram-tiktok.

Newcomb, Emma, Dominique Walk, Holland Haverkamp, et al. "Breaking down 'Harassment' to Characterize Trends in Human Interaction Cases in Maine's Pinnipeds." *Conservation Science and Practice* (October 1, 2021). https://doi.org/10.1111/csp2.518.

Patterson, Michael E., Jessica M. Montag, and Daniel R. Williams. "The Urbanization of Wildlife Management: Social Science, Conflict, and Decision Making." *Urban Forestry & Urban Greening* 1, no. 3 (2003): 171–83. https://doi.org/10.1078/1618-8667-00017.

Radnofsky, Caroline. "Freya the Walrus Delighted Norway. Her Death Has Divided the Country." NBC News (August 22, 2022). https://www.nbcnews.com/news/world/freya-walrus-euthanized-norway-outrage-rcna43073.

Rontziokos, Pamela. "Tasmania's Neil the Seal Has Found Viral Fame, Leaving Experts Concerned for His Welfare." *The Guardian* (December 21, 2023). https://www.theguardian.com/australia-news/2023/dec/22/neil-the-seal-tasmania-australia-outside-homes-footage-viral-fame-welfare.

Shepard, Paul. *The Others: How Animals Made Us Human*. Island Press, 1997.

United Press International. "Andre Back Home Following Week-Long Swim." *The Nassau Guardian* (May 4, 1985).

Yerbury, Rachel M., and Samantha J. Lukey. "Human–Animal Interactions: Expressions of Wellbeing through a 'Nature Language.'" *Animals* 11, no. 4 (2021): 950. https://doi.org/10.3390/ani11040950.

CHAPTER 3: THE SEAL BOUNTY CONSPIRACY

"About the Wabanaki Nations." Abbe Museum. https://www.abbemuseum.org/about-the-wabanaki-nations. Accessed September 13, 2023.

Bearchum, Jacob, Taylor Hensel, Adam Mazo, et al., co-directors. "Weckuwapok (The Approaching Dawn)." Video, Reciprocity Project (2022). https://www.reciprocity.org/films/weckuwapok.

Churchill, John "Al." "The 'Seal Nose' Scam." St. Croix Historical Society (May 10, 2022). https://stcroixhistorical.com/?p=3988.

Duff, John. "Public Shoreline Access in Maine: A Citizen's Guide to Ocean and Coastal Law, Third Edition." Maine Sea Grant College Program (August, 2016). https://www.maine.gov/dacf/parks/docs/public-shoreline-access-in-maine.pdf.

"Eastern Massachusetts." *The Springfield Weekly Republican* (February 20, 1908).

"Indians Arrive Under Arrest." *Boston Evening Transcript* (January 3, 1908).

Kalt, Joseph P., Amy Besaw Medford, and Jonathan B. Taylor. "Economic and Social Impacts of Restrictions on the Applicability of Federal Indian Policies to the Wabanaki Nations in Maine." Harvard Kennedy School (December 2022). https://ash.harvard.edu/wp-content/uploads/2024/02/wabanaki_report_vfin_for_dist_2022-12-09.pdf.

Lelli, Barbara, and David E. Harris. "Seal Bounty and Seal Protection Laws in Maine, 1872 to 1972: Historic Perspectives on a Current Controversy." *Natural Resources Journal* 46, no. 4 (2006): 881–924.

Lelli. "Seal Bounties in Maine and Massachusetts, 1888 to 1962."

"Made $15,000 by Bogus Seal Tails." *Daily Kennebec Journal* (January 13, 1908).

Maine Indian Program. *The Wabanakis of Maine and the Maritimes*. The Maine Indian Program of the New England Regional Office of the American Friends Service Committee (1989). https://files.eric.ed.gov/fulltext/ED393621.pdf.

"Maine Seals Believed to Have Passed for Bay State Product." *Daily Kennebec Journal* (November 26, 1907).

Maine Indian Tribal-State Commission. "Summary of the Maine Indian Land Claims Settlement of 1980." https://www.mitsc.org/mitsc-narrative-summaries/summary-of-the-maine-indian-land-claims-act-of-1980. Accessed January 5, 2024.

"Maine Indian Claims Settlement." Maine State Legislature. https://www.maine.gov/legis/lawlib/lldl/indianclaims/index.html. Accessed March 15, 2024.

Newsom, Bonnie D., Donald Soctomah, Emily Blackwood, and Jason A. Brough. "Indigenous Archaeologies, Shell Heaps, and Climate Change." *Advances in Archaeological Practice* 11, no. 3 (August 2023): 302–13. https://doi.org/10.1017/aap.2023.14.

Passamaquoddy Tribe at Indian Township. "Culture and History." https://www.passamaquoddy.com/?page_id=24. Accessed September 13, 2023.

Scully, Diana. "Maine Indian Claims Settlement: Concepts, Context, and Perspectives." Maine Indian Tribal-State Commission (February 14, 1995). https://digitalmaine.com/cgi/viewcontent.cgi?article=1010&context=mitsc_docs.

Sharp, David. "Maine Governor Vetoes Proposal Sought by Tribes to Ensure They Benefit from Federal Laws." Associated Press

(June 30, 2023). https://apnews.com/article/tribal-sovereignty-maine-tribes-federal-law-13ef67886d5509f1af8f0c8b39e7ff7a.

Sipayik Tribal Government. "Welcome | The Passamaquoddy: Peskotomuhkati." Accessed September 13, 2023. https://wabanaki.com.

"Socabasin Arrested." *The Boston Globe* (January 8, 1908).

Soctomah, Donald. *Passamaquoddy at the Turn of the Century: 1890-1920*. Passamaquoddy Tribe of Indian Township and Maine Humanities Council, 2002.

Soctomah, Donald. *Let Me Live as My Ancestors Had: 1850-1890*. Passamaquoddy Tribe of Indian Township and Maine Humanities Council, 2005.

Soctomah, Donald. "A Visit to Our Ancestors Place | Meddybemps - N'tolonapemk Village." Passamaquoddy Tribal Historic Preservation Office (2005). https://semspub.epa.gov/work/01/254079.pdf. Accessed September 13, 2023.

U.S. Congress. "H.R.7919—Maine Indian Claims Settlement Act of 1980." 96th Congress (September 22, 1980). https://www.congress.gov/bill/96th-congress/house-bill/7919.

Wabanaki Alliance. "Understanding Tribal Sovereignty." https://www.wabanakialliance.com/sovereignty. Accessed September 10, 2023.

CHAPTER 4: THE MIMIC

Beem, Heather R., and Michael S. Triantafyllou. "Wake-Induced 'Slaloming' Response Explains Exquisite Sensitivity of Seal Whisker-like Sensors." *Journal of Fluid Mechanics* 783 (October 2015): 306–22. https://doi.org/10.1017/jfm.2015.513.

Corkeron, Peter. "Captivity." In *Encyclopedia of Marine Mammals*, edited by William F. Perrin, Bernd Würsig, and J.G.M. Thewissen. Academic Press, 2009.

Duengen, Diandra, W. Tecumseh Fitch, and Andrea Ravignani. "Hoover the Talking Seal." *Current Biology* 33, no. 2 (2023): R50–52. https://doi.org/10.1016/j.cub.2022.12.023.

Hall. "Seals, Sea Lions, and Walruses."

Hannah, Janice. *Seals of Atlantic Canada and the Northeastern United States*. International Marine Mammal Association Inc., 2005.

Johnson, William, and David Lavigne. *Monk Seals in Antiquity*. Netherlands Commission for International Nature Protection, 1999.

Katona, Steven K., Valerie Rough, and David T. Richardson. *A Field Guide to the Whales, Porpoises, and Seals of the Gulf of Maine and Eastern Canada*. Macmillan Publishing Company, 1983.

Kienle, Sarah S., Roxanne D. Cuthbertson, and Joy S. Reidenberg. "Comparative Examination of Pinniped Craniofacial Musculature and Its Role in Aquatic Feeding." *Journal of Anatomy* 240, no. 2 (2021): 226–52. https://doi.org/10.1111/joa.13557.

Kim, Daniel. "'Super Senior' Barney the Seal Turns 38 at the Seattle Aquarium." *The Seattle Times* (September 14, 2023). https://www.seattletimes.com/seattle-news/super-senior-barney-the-seal-turns-38-at-the-seattle-aquarium.

Kleinfield, N. R. "Farewell to Gus, Whose Issues Made Him a Star." *The New York Times* (August 28, 2013). https://www.nytimes.com/2013/08/29/nyregion/gus-new-yorks-most-famous-polar-bear-dies-at-27.html.

McCulloch, S. and D. J. Boness. "Mother–Pup Vocal Recognition in the Grey Seal (*Halichoerus grypus*) of Sable Island, Nova Scotia." *Journal of Zoology* 251 (February 28, 2006): 449–455. https://doi.org/10.1111/j.1469-7998.2000.tb00800.x.

Moore, Bruce. "The Evolution of Imitative Learning." In *Social Learning in Animals*, edited by Cecilia Hayes and Bennett Galef. Academic Press, 1996.

Niesterok, Benedikt, Yvonne Krüger, Sven Wieskotten, Guido Dehnhardt, and Wolf Hanke. 2017. "Hydrodynamic Detection and Localization of Artificial Flatfish Breathing Currents by Harbour Seals (*Phoca vitulina*)." *The Journal of Experimental Biology* 220, no. 2 (2017): 174–85. https://doi.org/10.1242/jeb.148676.

NOAA Fisheries. "Kitt the Harbor Seal Has Died." February 2, 2023. Last modified March 22, 2023. https://www.fisheries.noaa.gov/feature-story/kitt-harbor-seal-has-died.

"Oldest Harbor Seal on Record Turns 48!" Oregon Coast Aquarium. https://aquarium.org/skinny-turns-48. Accessed December 15, 2022.

"Seals' Whiskers Provide a Model for the Latest Submarine Detectors." *The Economist* (August 9, 2018). https://www.economist.com/science-and-technology/2018/08/09/seals-whiskers-provide-a-model-for-the-latest-submarine-detectors.

St. John, Susan. *To Sail Beyond the Sunset: Hurricane Island—The Outward Bound Years*. Philip Conkling & Associates, 2017.

Pliny the Elder. *The Natural History.* John Bostock, M.D., F.R.S. H.T. Riley, Esq., B.A. London. Taylor and Francis, Red Lion Court, Fleet Street, 1855.

CHAPTER 5: WHY DID THE SEAL CROSS THE ROAD?

Estes, James A., Michael Heithaus, Douglas J. McCauley, Douglas B. Rasher, and Boris Worm. "Megafaunal Impacts on Structure and Function of Ocean Ecosystems." *Annual Review of Environment and Resources* 41, no. 1 (October 2016): 83–116. https://doi.org/10.1146/annurev-environ-110615-085622.

Estes, James A., Robert S. Steneck, and David R. Lindberg. "Exploring the Consequences of Species Interactions through the Assembly and Disassembly of Food Webs: A Pacific-Atlantic Comparison." *Bulletin of Marine Science* 89, no. 1 (January 2013): 11–29. https://doi.org/10.5343/bms.2011.1122.

Kelaher, Brendan P., Mélissa Tan, Will F. Figueira, et al. "Fur Seal Activity Moderates the Effects of an Australian Marine Sanctuary on Temperate Reef Fish." *Biological Conservation* 182 (February, 2015): 205–14. https://doi.org/10.1016/j.biocon.2014.12.011.

Link, J. "Does Food Web Theory Work for Marine Ecosystems?" *Marine Ecology Progress Series* 230 (April 2002): 1–9. https://doi.org/10.3354/meps230001.

Matthiessen, Peter. *Wildlife in America.* Viking, 1959.

Morris, Tina. *Return to the Sky.* Chelsea Green Publishing, 2024.

"Northern Sea Otter." Marine Mammal Commission. https://www.mmc.gov/priority-topics/species-of-concern/northern-sea-otters. Accessed February 27, 2023.

Raymond, Wendel W., Brent B. Hughes, Tiffany A. Stephens, Catherine R. Mattson, Ashley T. Bolwerk, and Ginny L. Eckert. "Testing the Generality of Sea Otter–Mediated Trophic Cascades in Seagrass Meadows." *Oikos* 130, no. 5 (2021): 725–38. https://doi.org/10.1111/oik.07681.

Rough, Valerie. "Gray Seals in Nantucket Sound, Massachusetts, Winter and Spring, 1994: Final Report to the Marine Mammal Commission." U.S. Marine Mammal Commission (March 1995).

Wood. "Rates of Increase in Gray Seal (*Halichoerus grypus atlantica*) Pupping at Recolonized Sites in the United States, 1988–2019."

Wood, Stephanie A., Elizabeth Josephson, Kristin Precoda, and Kimberly T. Murray. "Gray Seal (*Halichoerus grypus*) Pupping Trends and 2021 Population Estimate in U.S. Waters." Northeast Fisheries Science Center Reference Document (2022): 22–14.

CHAPTER 6: THE GRAVEYARD OF THE ATLANTIC

Armstrong, Bruce. *Sable Island*. Formac, 2010.

Bowen, W. Don, Sara L. Ellis, Sara J. Iverson, and Daryl J. Boness. "Maternal and Newborn Life-History Traits during Periods of Contrasting Population Trends: Implications for Explaining the Decline of Harbour Seals (*Phoca vitulina*), on Sable Island." *Journal of Zoology* 261, no. 2 (2006): 155–63. https://doi.org/10.1017/s0952836903004047.

De Villiers, Marq, and Sheila Hirtle. *Sable Island: The Strange Origins and Curious History of a Dune Adrift in the Atlantic*. Walker & Company, 2004.

Devred, Emmanuel, Andrea Hilborn, and Cornelia Elizabeth den Heyer. "Enhanced Chlorophyll-*A* Concentration in the Wake of Sable Island, Eastern Canada, Revealed by Two Decades of Satellite Observations: A Response to Grey Seal Population Dynamics?" *Biogeosciences* 18, no. 23 (2021): 6115–32. https://doi.org/10.5194/bg-18-6115-2021.

Donovan, Moira. "Romance, Politics, and Ecological Damage: The Saga of Sable Island's Wild Horses." *Hakai Magazine* (July 26, 2022). https://hakaimagazine.com/features/romance-politics-and-ecological-damage-the-saga-of-sable-islands-wild-horses.

Freedman, Bill. *Sable Island: The Ecology and Biodiversity of Sable Island*. Fitzhenry and Whiteside, 2016.

Cousteau, Jacques-Yves, and Jacques Gagné. "St. Lawrence: Stairway to the Sea." Video, National Film Board of Canada (1982).

Homer. *The Odyssey*. Translated by Emily Wilson. W. W. Norton & Company, 2018.

Lambert. "The Grey Seal in Britain."

McLoughlin, Philip D., Kenton Lysak, Lucie Debeffe, Thomas Perry, and Keith A. Hobson. "Density-Dependent Resource Selection by a Terrestrial Herbivore in Response to Sea-to-Land Nutrient Transfer by Seals." *Ecology* 97, no. 8 (2016): 1,929–37. https://doi.org/10.1002/ecy.1451.

Morris, James Rainstorpe, and Rosalee Stilwell. *Sable Island Journals 1801–1804*. Sable Island Preservation Trust, 2001.

"Odyssey Sirens 'Were Monk Seals.'" BBC News (May 19, 2005). http://news.bbc.co.uk/1/hi/world/europe/4559217.stm.

Pitman, Robert L., John A. Totterdell, Holly Fearnbach, Lisa T. Ballance, John W. Durban, and Hans Kemps. "Whale Killers: Prevalence and Ecological Implications of Killer Whale Predation on Humpback Whale Calves off Western Australia." *Marine Mammal Science* 31, no. 2 (2014): 629–57. https://doi.org/10.1111/mms.12182.

Roman, Joe. *Eat, Poop, Die*. Little, Brown Spark, 2023.

Wood, S. A., T. R. Frasier, B. A. McLeod, et al. "The Genetics of Recolonization: An Analysis of the Stock Structure of Grey Seals (*Halichoerus grypus*) in the Northwest Atlantic." *Canadian Journal of Zoology* 89, no. 6 (2011): 490–97. https://doi.org/10.1139/z11-012.

CHAPTER 7: THE CORKSCREW SEAL MYSTERY

Bishop, Amanda M., Joseph Onoufriou, Simon Moss, Paddy P. Pomeroy, and Sean D. Twiss. "Cannibalism by a Male Grey Seal (*Halichoerus grypus*) in the North Sea." *Aquatic Mammals* 42, no. 2 (2016): 137–43. https://doi.org/10.1578/am.42.2.2016.137.

Bodin, Madeline. "Mystery of the Corkscrew Seals." *bioGraphic* (May 24, 2016). https://www.biographic.com/mystery-of-the-corkscrew-seals.

Bowen, W. D. and G. D. Harrison. "Seasonal and Interannual Variability in Grey Seal Diets on Sable Island, Eastern Scotian Shelf." *NAMMCO Scientific Publications* 6 (2007): 123–134. https://doi.org/10.7557/3.2728.

Bowen, W. D., J. I. McMillan, and Wade Blanchard. "Reduced Population Growth of Gray Seals at Sable Island: Evidence from Pup Production and Age of Primiparity." *Marine Mammal Science* 23, no. 1 (January 2007): 48–64. https://doi.org/10.1111/j.1748-7692.2006.00085.x.

Brownlow, Andrew, Joseph Onoufriou, Amanda Bishop, Nicholas Davison, and Dave Thompson. "Corkscrew Seals: Grey Seal (*Halichoerus grypus*) Infanticide and Cannibalism May Indicate the Cause of Spiral Lacerations in Seals." *PLOS ONE* 11, no. 6 (2016): e0156464. https://doi.org/10.1371/journal.pone.0156464.

den Heyer, C. E., W. D. Bowen, and J. I. McMillan. "Long-term Changes in Grey Seal Vital Rates at Sable Island Estimated from POPAN Mark-Resighting Analysis of Branded Seals." DFO Canadian Science Advisory Secretariat. Res. Doc. 2013/021.

den Heyer, Cornelia E., W. Don Bowen, Julian Dale, et al. "Contrasting Trends in Gray Seal (*Halichoerus grypus*) Pup Production throughout the Increasing Northwest Atlantic Metapopulation." *Marine Mammal Science* 37, no. 2 (2020): 611–30. https://doi.org/10.1111/mms.12773.

de Waal, Frans. "Are We in Anthropodenial?" *Discover Magazine* (January 18, 1997). https://www.discovermagazine.com/planet-earth/are-we-in-anthropodenial.

Freedman. *Sable Island*.

Lidgard, Damian C., Daryl J. Boness, W. Don Bowen, and Jim I. McMillan. "State-Dependent Male Mating Tactics in the Grey Seal: The Importance of Body Size." *Behavioral Ecology* 16, no. 3 (2005): 541–549. https://doi.org/10.1093/beheco/ari023.

Lucas, Zoe N., and Lisa J. Natanson. "Two Shark Species Involved in Predation on Seals at Sable Island, Nova Scotia, Canada." *Proceedings of the Nova Scotian Institute of Science* 45, no. 2 (2010): 64–88.

Lucas, Z. and W. T. Stobo. "Shark-Inflicted Mortality on a Population of Harbour Seals (*Phoca vitulina*) at Sable Island, Nova Scotia." *Journal of Zoology* 252 (2000): 405–414. https://doi.org/10.1111/j.1469-7998.2000.tb00636.x.

Nicoletti, Angela. "Mysterious Arctic Shark Spotted in the Caribbean Thousands of Miles from Home." Florida International University News (August 18, 2024). https://news.fiu.edu/2022/greenland-shark.

North Atlantic Marine Mammal Commission. "Hooded Seal." Last modified July 2021. https://nammco.no/hooded-seal/#1475762140594-0925dd6e-f6cc.

van Neer, Abbo, Stephanie Gross, Tina Kesselring, et al. "Assessing Seal Carcasses Potentially Subjected to Grey Seal Predation." *Scientific Reports* 11, no. 1 (2021). https://doi.org/10.1038/s41598-020-80737-9.

Watanabe, Yuuki Y., Christian Lydersen, Aaron T. Fisk, and Kit M. Kovacs. 2012. "The Slowest Fish: Swim Speed and Tail-Beat Frequency of Greenland Sharks." *Journal of Experimental Marine Biology and Ecology* 426–427 (September 2012): 5–11. https://doi.org/10.1016/j.jembe.2012.04.021.

CHAPTER 8: SCAMPS OR SCAPEGOATS?

Ampela, Kristen. "The Diet and Foraging Ecology of Gray Seals (*Halichoerus grypus*) in United States Waters." PhD diss., The City University of New York (2009). http://doi.org/10.13140/RG.2.2.17950.46400.

Arnaquq-Baril, Alethea. "Angry Inuk." Video, National Film Board of Canada (2016).

Bowen, W. D., J. W. Lawson, and B. Beck. "Seasonal and Geographic Variation in Species Composition and Size of Prey Consumed by Grey Seals (*Halichoerus grypus*) on the Scotian Shelf." *Canadian Journal of Fisheries and Aquatic Science* 50 (1993): 1,768–1,778.

Brown, Cassie, and Harold Horwood. *Death on the Ice: the Great Newfoundland Sealing Disaster of 1914*. Anchor Canada, 2016.

Chase, Chris. "As Canada Considers Seal Cull, Fisheries Representatives Urge Caution." Seafood Source (March 29, 2023). https://www.seafoodsource.com/news/environment-sustainability/as-canada-considers-seal-cull-fisheries-representatives-urge-caution.

Clayton, Mark. "As Cod Industry Wanes, Canada Announces Controversial Seal Hunt." *The Christian Science Monitor* (November 14, 1995). https://www.csmonitor.com/1995/1114/14102.html.

Cochrane, Candace, Andrea Procter, and Nunatsiavut Creative Group. *Tautukkonik | Looking Back*. Memorial University Press, 2022.

Cochrane, Candace. *Outport*. Flanker Press, 2008.

Fink, Sheryl. "Why Hasn't Canada Ended Its Commercial Seal Hunt?" International Fund for Animal Welfare (April 6, 2023). https://www.ifaw.org/journal/no-end-canada-seal-hunt.

Frum, Barbara. "CBC Interview of Paul Watson." Audio, CBC Radio (1978).

Hernandez, K. M., A. L. Bogomolni, J. H. Moxley, et al. "Seasonal Variability and Individual Consistency in Gray Seal (*Halichoerus grypus*) Isotopic Niches." *Canadian Journal of Zoology* 97, no. 11 (2019): 1071–77. https://doi.org/10.1139/cjz-2019-0032.

Holland, Eva. "An Act of Forgiveness Fuels a Fight in the Arctic." *Hakai Magazine* (November 29, 2016). https://hakaimagazine.com/features/act-forgiveness-fuels-fight-arctic.

Kurlansky, Mark. *Cod: A Biography of the Fish That Changed the World.* Penguin Books, 1998.

Lister-Kaye, John. *Seal Cull.* Penguin Group, 1979.

Lyssikatos, Marjorie C., and Frederick W. Wenzel. "What Bycatch Tells Us about the Diet of Harbor and Gray Seals and Overlap with Commercial Fishermen." *Frontiers in Conservation Science* 5 (April 2024). https://doi.org/10.3389/fcosc.2024.1377673.

Mackenzie, Debora. "Seals to the Slaughter." *New Scientist* (March 16, 1996). https://www.newscientist.com/article/mg14920213-900-seals-to-the-slaughter.

McCosker, Christina M., Zachary H. Olson, and Kathryn A. Ono. "A Comparative Methodological Approach to Studying the Diet of a Recovering Marine Predator, the Grey Seal (*Halichoerus grypus*)." *Canadian Journal of Zoology* 102, no. 2 (2024): 182–94. https://doi.org/10.1139/cjz-2023-0104.

Pannozzo, Linda. "How to Kill 220,000 Seals on Sable Island: The DFO Plan." *The Coast* (May 27, 2010). https://www.thecoast.ca/news-opinion/how-to-kill-220000-seals-on-sable-island-the-dfo-plan-1663224.

Pannozzo, Linda. *The Devil and the Deep Blue Sea.* Fernwood Publishing, 2013.

Pelly, David F. *Sacred Hunt.* University of Washington Press, 2001.

Pershing, Andrew J., Michael A. Alexander, Christina M. Hernandez, et al. "Slow Adaptation in the Face of Rapid Warming Leads to Collapse of the Gulf of Maine Cod Fishery." *Science* 350, no. 6262 (2015): 809–12. https://doi.org/10.1126/science.aac9819.

Rosano, Michela. "Cod Moratorium: How Newfoundland's Cod Industry Disappeared Overnight." *Canadian Geographic* (July 11, 2022). https://canadiangeographic.ca/articles/cod-moratorium-how-newfoundlands-cod-industry-disappeared-overnight.

Talbot, Lee. "Congress Is Willing to Risk Lives of Dwindling Whales and Dolphins." *The Hill* (November 14, 2017). https://thehill.com/opinion/energy-environment/360294-congress-cant-seriously-consider-rolling-back-protections-for.

Verma, Jenn Thornhill. "Scapegoat or Scoundrel? Why Scientists Want to Clear the Air about the Role of Seals and Focus on Ecosystems." *The Globe and Mail* (August 29, 2021). https://www.theglobeandmail.com

/canada/article-scapegoat-or-scoundrel-why-scientists-want-to-clear-the-air-about-the.

Verma, Jenn. "The Cod Delusion." *Canadian Geographic* (July 1, 2022). https://canadiangeographic.ca/articles/the-cod-delusion.

Whitehead, Hal, Sara Iverson, Boris Worm, and Heike Lotze. "Independent Marine Scientists Respond to Senate Fisheries Committee Report 'The Sustainable Management of Grey Seal Populations: A Path toward the Recovery of Cod and Other Groundfish Stocks.'" GlobeNewswire news release (November 6, 2012). https://www.globenewswire.com/news-release/2012/11/06/1425132/0/en/Independent-Marine-Scientists-Respond-to-Senate-Fisheries-Committee-Report-The-Sustainable-Management-of-Grey-Seal-Populations-A-Path-Toward-the-Recovery-of-Cod-and-Other-Groundfis.html.

CHAPTER 9: BLUBBER BUSTERS

Brown, Matthew, and John Flesher. "Most Money for Endangered Species Goes to a Small Number of Creatures, Leaving Others in Limbo." Associated Press (December 30, 2023). https://apnews.com/article/endangered-species-spending-extinctions-plants-1ad806deodb9d09a38b7e82f6286c1b5.

"Columbia River Sea Lion Management." Washington Department of Fish & Wildlife. https://wdfw.wa.gov/species-habitats/at-risk/species-recovery/sea-lion-management. Accessed January 3, 2024.

"Environmental Assessment on Conditions for Lethal Removal of California Sea Lions at the Ballard Locks to Protect Winter Steelhead." U.S. Department of Commerce (1996). March Report.

"Federal and State Endangered and Threatened Species Expenditures." U.S. Fish & Wildlife Service (2020). https://www.fws.gov/sites/default/files/documents/endangered-threatened-species-expenditures-%20report-to-congress-fiscal-year-2020.pdf.

Fazio, Marie. "Northwest's Salmon Population May Be Running Out of Time." *The New York Times* (January 20, 2021). https://www.nytimes.com/2021/01/20/climate/washington-salmon-extinction-climate-change.html.

Gammon, Kate. "Herschel, the Very Hungry Sea Lion." *Hakai Magazine* (October 16, 2018). https://hakaimagazine.com/features/herschel-the-very-hungry-sea-lion.

Goldfarb, Ben. "For Sea Lions, a Feast of Salmon on the Columbia." *High Country News* (July 6, 2015). https://www.hcn.org/articles/on-the-columbia-river-what-do-you-do-with-a-hungry-sea-lion.

Jeffries, Steven J. and Joe Scordino. "Efforts to Protect a Winter Steelhead Run from California Sea Lion Predation at the Ballard Locks." Pinniped Populations, Eastern North Pacific: Status, Trends and Issues, (August 28, 1997).

Levin, Josh. "One Year: 1986, Herschel vs. the Blubber Busters." Podcast, *Slate* (September 15, 2022).

NOAA Fisheries. "California Sea Lion." Last modified April 21, 2022. https://www.fisheries.noaa.gov/species/california-sea-lion.

NOAA Fisheries. "Sea Lion Removals Annual Summary." Last modified March 18, 2024. https://www.fisheries.noaa.gov/west-coast/marine-mammal-protection/sea-lion-removals-annual-summary.

Pearson, Scott F., Staci M. Amburgey, Casey T. Clark, et al. "Trends and Status of Harbor Seals in Washington State, USA (1977–2023)." *Marine Mammal Science* (2024). https://doi.org/10.1111/mms.13161.

U.S. Congress. "Marine Mammal Protection Act."

"National Marine Fisheries Service OK's Permanent Home for Seattle's Sea Lions at Sea World in Florida." U.S. Department of Commerce. Media Release (April 4, 1996). Courtesy of Joe Scordino.

U.S. Department of Commerce. "Impacts of California Sea Lions and Pacific Harbor Seals on Salmonids and West Coast Ecosystems." NOAA Fisheries, Report to Congress (February 10, 1999).

The White House. "Columbia River Basin Fisheries: Working Together to Develop a Path Forward." Last modified May 4, 2023. https://bidenwhitehouse.archives.gov/ceq/news-updates/2022/03/28/columbia-river-basin-fisheries-working-together-to-develop-a-path-forward/.

CHAPTER 10: THE FISH WARS

Dougherty, Phil. "Boldt Decision: *United States v. State of Washington*." Historylink.org (August 24, 2020). https://www.historylink.org/file/21084.

Dwyer, Colin. "After Calf's Death, Orca Mother Carries It for Days in 'Tragic Tour of Grief.'" NPR (July 31, 2018). https://www.npr.org/2018

/07/31/6343147141/after-calfs-death-orca-mother-carries-it-for-days-in-tragic-tour-of-grief.

Egan, Timothy. "On Good Days, the Smell Can Hardly Be Noticed." *The New York Times* (April 6, 1988). https://www.nytimes.com/1988/04/06/us/tacoma-journal-on-good-days-the-smell-can-hardly-be-noticed.html.

"The Fish Wars." Smithsonian National Museum of the American Indian (2018). https://americanindian.si.edu/nk360/pnw-fish-wars/#title. Accessed May 5, 2024.

"Indian and Tribal Law." University of Washington School of Law. 2024. Last modified July 31. https://lib.law.uw.edu/c.php?g=1239321&p=9069754.

McGrath, Maggie. "The Age of Disruption: Meet the 50 Over 50 2023: Profile Ramona Bennett." *Forbes* (August 1, 2023). https://www.forbes.com/profile/ramona-bennett.

Pailthorp, Bellamy. "Seals and Sea Lions Vex Washington Tribes as Marine Mammal Protection Act Turns 50." KNKX Public Radio (November 3, 2022). https://www.knkx.org/environment/2022-11-03/seals-and-sea-lions-vex-washington-tribes-as-marine-mammal-protection-act-turns-50.

Pailthorp, Bellamy. "The Standoff at This Pierce County Bridge 50 Years Ago Codified Tribal Treaty Fishing Rights." KNKX Public Radio (December 29, 2020). https://www.knkx.org/environment/2020-12-29/the-standoff-at-this-pierce-county-bridge-50-years-ago-codified-tribal-treaty-fishing-rights.

"About Our Tribe." Puyallup Tribe of Indians. https://www.puyalluptribe-nsn.gov/about-our-tribe. Accessed May 5, 2024.

Schick, Tony, and Irena Hwang. "PNW Hatcheries Aren't Saving Salmon, Investigation Finds." Oregon Public Broadcasting and ProPublica (July 1, 2022). https://www.cascadepbs.org/environment/2022/07/pnw-hatcheries-arent-saving-salmon-investigation-finds.

Stehr, Carla M., Donald W. Brown, Tom Hom, Bernadita F. Anulacion, William L. Reichert, and Tracy K. Collier. "Exposure of Juvenile Chinook and Chum Salmon to Chemical Contaminants in the Hylebos Waterway of Commencement Bay, Tacoma, Washington." *Journal of Aquatic Ecosystem Stress and Recovery* 7, no. 3 (2000): 215–27. https://doi.org/10.1023/a:1009905322386.

"Understanding Tribal Treaty Rights in Western Washington." Northwest Indian Fisheries Commission. https://nwifc.org/w/wp-content/uploads

/downloads/2014/10/understanding-treaty-rights-final.pdf. Accessed April 3, 2023.

"United States v. Washington, 384 F. Supp. 312 (1974)." United States District Court for the Western District of Washington (1974).

CHAPTER 11: THE CHASE

Bogomolni, Andrea, Owen C. Nichols, and Dee Allen. "A Community Science Approach to Conservation Challenges Posed by Rebounding Marine Mammal Populations: Seal-Fishery Interactions in New England." *Frontiers in Conservation Science* 2 (2021): 696535. https://doi.org/10.3389/fcosc.2021.696535.

Freedman, Ethan. "Seals Show Scientists an Unknown Antarctic Canyon." *Scientific American* (November 1, 2023). https://www.scientificamerican.com/article/seals-show-scientists-an-unknown-antarctic-canyon.

Lyssikatos. "What Bycatch Tells Us about the Diet of Harbor and Gray Seals and Overlap with Commercial Fishermen."

McMahon, Clive R., Mark A. Hindell, Jean Benoit Charrassin, et al. "Southern Ocean Pinnipeds Provide Bathymetric Insights on the East Antarctic Continental Shelf." *Communications Earth & Environment* 4, no. 1 (2023): 1–10. https://doi.org/10.1038/s43247-023-00928-w.

Murray, K. T., J. M. Hatch, R. A. DiGiovanni, and E. Josephson. "Tracking Young-of-the-Year Gray Seals *Halichoerus grypus* to Estimate Fishery Encounter Risk." *Marine Ecology Progress Series* 671 (August 2021): 235–45. https://doi.org/10.3354/meps13765.

Nichols, Owen C., Ernie Eldredge, and Steven X. Cadrin. "Gray Seal Behavior in a Fish Weir Observed Using Dual-Frequency Identification Sonar." *Marine Technology Society Journal* 48, no. 4 (2014): 72–78. https://doi.org/10.4031/mtsj.48.4.2.

Nichols, O. C., A. Bogomolni, E. Bradfield, G. Early, L. Sette, and S. Wood. "Gulf of Maine Seals—Fisheries Interactions and Integrated Research: Final Report." Woods Hole Oceanographic Institution Technical Report (2012). https://hdl.handle.net/1912/5514.

Puryear, Wendy, Kaitlin Sawatzki, Nichola Hill, et al. "Highly Pathogenic Avian Influenza A(H5N1) Virus Outbreak in New England Seals, United States." *Emerging Infectious Diseases* 29, no. 4 (2023): 786–791. https://doi.org/10.3201/eid2904.221538.

CHAPTER 12: THE SEAL SNATCHER

Acquarone, Mario. "Harbour Seals in the North Atlantic and the Baltic." *NAMMCO Scientific Publications* 8, no. 3 (2010).

Morris, Alix. "Seal Populations Are Booming, and So Are Rescues by Marine Mammals of Maine." *Down East Magazine* (November, 2023). https://downeast.com/land-wildlife/seal-populations-are-booming-and-so-are-rescues-by-marine-mammals-of-maine.

Newcomb. "Breaking Down 'Harassment' to Characterize Trends in Human Interaction Cases in Maine's Pinnipeds."

CHAPTER 13: PINNIPEDS AS PREY

Behnke, Jim. "Great White Sharks Are Surging off Cape Cod." *Scientific American* (July 2, 2023). https://www.scientificamerican.com/article/remarkable-rebound-the-great-white-sharks-of-cape-cod1.

Benchley, Peter. *Jaws*. Doubleday, 1974.

Blei, Daniela. "Inventing the Beach: The Unnatural History of a Natural Place." *Smithsonian* (June 23, 2016). https://www.smithsonianmag.com/history/inventing-beach-unnatural-history-natural-place-180959538.

Castro, Jose. "The Origins and Rise of Shark Biology in the 20th Century." *Marine Fisheries Review* 78, no. 1–2 (2017): 14–33.

Corbin, Alain. *The Lure of the Sea: The Discovery of the Seaside in the Western World, 1750–1840*. University Of California Press, 1994.

"How Jaws Misrepresented the Great White." BBC News (June 8, 2015). https://www.bbc.com/news/magazine-33049099.

Mollomo, Paul. "The White Shark in Maine and Canadian Atlantic Waters." *Northeastern Naturalist* 5, no. 3 (1998): 207–214. https://doi.org/10.2307/3858620.

"Population and Housing." National Ocean Economics Program. 2017. Last modified September 20. https://www.oceaneconomics.org/NOEP/Demographics.

Skomal, Greg, and Ret Talbot. *Chasing Shadows*. HarperCollins, 2023.

Skomal, Greg, John Chisholm, and Steven Correia. "Implications of Increasing Pinniped Populations on the Diet and Abundance of White Sharks off the Coast of Massachusetts." In *Global Perspectives on the*

Biology and Life History of the White Shark, edited by Michael L. Domeier. CRC Press, 2012.

Swanson, Ana. "The Weird Origins of Going to the Beach." *Washington Post* (July 3, 2016). https://www.washingtonpost.com/news/wonk/wp/2016/07/03/the-weird-origins-of-going-to-the-beach.

Thoreau, Henry David. *Cape Cod*. Ticknor and Fields, 1865.

Wilkinson, Alec. "A Deadly Shark Attack at a Beach on Cape Cod That I Know Well." *The New Yorker* (September 17, 2018). https://www.newyorker.com/culture/culture-desk/a-deadly-shark-attack-at-a-beach-on-cape-cod-that-i-know-well.

Winton, Megan V., James Sulikowski, and Gregory B. Skomal. "Fine-Scale Vertical Habitat Use of White Sharks at an Emerging Aggregation Site and Implications for Public Safety." *Wildlife Research* 48, no. 4 (2021): 345. https://doi.org/10.1071/wr20029.

Winton, Megan V., Gavin Fay, and Gregory B. Skomal. "An Open Spatial Capture-Recapture Framework for Estimating the Abundance and Seasonal Dynamics of White Sharks at Aggregation Sites." *Marine Ecology Progress Series* 715 (July 2023): 1–25. https://doi.org/10.3354/meps14371.

"Outer Cape Shark Mitigation Alternatives Analysis." Woods Hole Group (October 2019). https://www.scribd.com/document/430571977/Shark-Mitigation-Alternatives-Analysis-Technical-Report-10112019.

CHAPTER 14: WILD CAPE COD

Bogomolni. "A Community Science Approach to Conservation Challenges Posed by Rebounding Marine Mammal Populations."

Bratton, Rachel, Jennifer L. Jackman, Stephanie A. Wood, et al. "Conflict with Rebounding Populations of Marine Predators: Management Preferences of Three Stakeholder Groups on Cape Cod, Massachusetts." *Ocean & Coastal Management* 244, no. 6 (2023): 106800. https://doi.org/10.1016/j.ocecoaman.2023.106800.

Cammen. "Predator Recovery, Shifting Baselines, and the Adaptive Management Challenges They Create."

Howell, Peter. "Gray Seals No Longer Need Federal Protection." *Boston Globe Opinion* (August 1, 2023). https://www.bostonglobe.com/2023/08/01/opinion/gray-seals-sharks-marine-mammal-protection-act.

Jackman, Jennifer, Lauren Bettencourt, Jerry Vaske, et al. "Conflict and Consensus in Stakeholder Views of Seal Management on Nantucket Island, MA, USA." *Marine Policy* 95 (2018): 166–73. https://doi.org/10.1016/j.marpol.2018.03.006.

Bratton, Rachel, Jennifer L Jackman, Stephanie A Wood, et al. "Conflict with Rebounding Populations of Marine Predators: Management Preferences of Three Stakeholder Groups on Cape Cod, Massachusetts." *Ocean & Coastal Management* 244 (2023): 106800. https://doi.org/10.1016/j.ocecoaman.2023.106800.

Jackman, Jennifer L., Jerry J. Vaske, Seana Dowling-Guyer, Rachel Bratton, Andrea L. Bogomolni, and Stephanie A. Wood. "Seals and the Marine Ecosystem: Attitudes, Ecological Benefits/Risks and Lethal Management Views." *Human Dimensions of Wildlife* 29, no. 3 (2023): 1–17. https://doi.org/10.1080/10871209.2023.2212686.

Pauly, Daniel. "Anecdotes and the Shifting Baseline Syndrome of Fisheries." *Trends in Ecology & Evolution* 10, no. 10 (1995): 430. https://doi.org/10.1016/s0169-5347(00)89171-5.

Roman, Joe, Meagan M. Dunphy-Daly, David W. Johnston, and Andrew J. Read. "Lifting Baselines to Address the Consequences of Conservation Success." *Trends in Ecology & Evolution* 30, no. 6 (2015): 299–302. https://doi.org/10.1016/j.tree.2015.04.003.

CHAPTER 15: WAR AND PEACEBUILDING

Butler, James R. A., Stuart J. Middlemas, Simon A. McKelvey, et al. "The Moray Firth Seal Management Plan: An Adaptive Framework for Balancing the Conservation of Seals, Salmon, Fisheries and Wildlife Tourism in the UK." *Aquatic Conservation: Marine and Freshwater Ecosystems* 18, no. 6 (2008): 1025–38. https://doi.org/10.1002/aqc.923.

Calfas, Jennifer. "Wolf Resurgence in Washington State Tests Limits of Civility." *The Wall Street Journal* (October 12, 2019). https://www.wsj.com/articles/wolf-resurgence-in-washington-state-tests-limits-of-civility-11570885202.

Cammen. "Predator Recovery, Shifting Baselines, and the Adaptive Management Challenges They Create."

Cammen, Kristina M., Thomas F. Schultz, W. Don Bowen, et al. "Genomic Signatures of Population Bottleneck and Recovery in Northwest Atlantic Pinnipeds." *Ecology and Evolution* 8, no. 13 (2018): 6,599–6,614. https://doi.org/10.1002/ece3.4143.

Cammen, Kristina M., Sarah Vincze, A. Sky Heller, et al. "Genetic Diversity from Pre-Bottleneck to Recovery in Two Sympatric Pinniped Species in the Northwest Atlantic." *Conservation Genetics* 19, no. 3 (2018): 555–69. https://doi.org/10.1007/s10592-017-1032-9.

Farquhar, Brodie. 2023. "Wolf Reintroduction Changes Ecosystem in Yellowstone." Yellowstone National Park (June 22, 2023). https://www.yellowstonepark.com/things-to-do/wildlife/wolf-reintroduction-changes-ecosystem.

Hoelzel, A. Rus, Georgios A. Gkafas, Hui Kang, et al. "Genomics of Post-Bottleneck Recovery in the Northern Elephant Seal." *Nature Ecology & Evolution* 8 (2024): 686-694. https://doi.org/10.1038/s41559-024-02337-4.

Ma, Michelle. "New Tool Maps a Key Food Source for Grizzly Bears: Huckleberries." University of Washington News (March 26, 2019). https://www.washington.edu/news/2019/03/26/new-tool-maps-a-key-food-source-for-grizzly-bears-huckleberries.

Madden, Francine, and Brian McQuinn. "Conservation's Blind Spot: The Case for Conflict Transformation in Wildlife Conservation." *Biological Conservation* 178, no. 4 (2014): 97–106. https://doi.org/10.1016/j.biocon.2014.07.015.

McKean, Andrew. "As Wolf Management Debate Reaches a Fever Pitch, the Interior Department Hires a National Mediator." *Outdoor Life* (April 15, 2024). https://www.outdoorlife.com/hunting/wolf-management-mediator.

Ripple, William J., Eric J. Larsen, Roy A. Renkin, and Douglas W. Smith. "Trophic Cascades among Wolves, Elk and Aspen on Yellowstone National Park's Northern Range." *Biological Conservation* 102, no. 3 (2001): 227–34. https://doi.org/10.1016/s0006-3207(01)00107-0.

Shores, Carolyn R., Nate Mikle, and Tab A. Graves. "Mapping a Keystone Shrub Species, Huckleberry (*Vaccinium membranaceum*), Using Seasonal Colour Change in the Rocky Mountains." *International Journal of Remote Sensing* 40, no. 15 (2019): 5695–5715. https://doi.org/10.1080/01431161.2019.1580819.

van Eeden, Lily M., Chris R. Dickman, Euan G. Ritchie, and Thomas M. Newsome. "Shifting Public Values and What They Mean for Increasing Democracy in Wildlife Management Decisions." *Biodiversity and Conservation* 26, no. 11 (2017): 2759–63. https://doi.org/10.1007/s10531-017-1378-9.

CHAPTER 16: THE SHAPE-SHIFTER

Eisley, Loren. *The Unexpected Universe*. Mariner Books, 1972.

Estes. "Megafaunal Impacts on Structure and Function of Ocean Ecosystems."

Jackson, Maggie. *Uncertain: The Wisdom and Wonder of Being Unsure*. Prometheus, 2023.

"Legend of Kópakonan (Seal Woman)." Visit Faroe Islands. https://old.visitfaroeislands.com/en/be-inspired/in-depth-articles/legend-of-kopakonan-(seal-woman). Accessed May 5, 2024.

"Marine Mammal Protection Act." Marine Mammal Commission. https://www.mmc.gov/about-the-commission/our-mission/marine-mammal-protection-act. Accessed July 27, 2023.

McCloskey, Robert. *One Morning in Maine*. The Viking Press, 1952.

Mikkelsen, Bjarni. "Present Knowledge of Grey Seals (*Halichoerus grypus*) in Faroese Waters." *NAMMCO Scientific Publications* 6 (January 2007): 79. https://doi.org/10.7557/3.2724.

Tróndarson, Sveinur. "Kópakonan Travel Guide." Guide to Faroe Islands (October 18, 2024). https://guidetofaroeislands.fo/travel-faroe-islands/drive/kopakonan.

Wilson, Edward O. *The Meaning of Human Existence*. Liveright, 2014.